Elias Colbert

Star Studies.

What we Know of the Universe outside the Earth

Elias Colbert

Star Studies.
What we Know of the Universe outside the Earth

ISBN/EAN: 9783337042356

Printed in Europe, USA, Canada, Australia, Japan

Cover: Foto ©berggeist007 / pixelio.de

More available books at **www.hansebooks.com**

Star-Studies.

WHAT WE KNOW

OF

OUTSIDE THE EARTH.

BY ELIAS COLBERT,

OF THE CHICAGO TRIBUNE,

Emeritus Assistant Director Dearborn Observatory;
Author of "Astronomy Without a Telescope."

PRICE, FIFTY CENTS.

CHICAGO:
THE WESTERN NEWS COMPANY.
1871.

The Lakeside Press.

Bind in

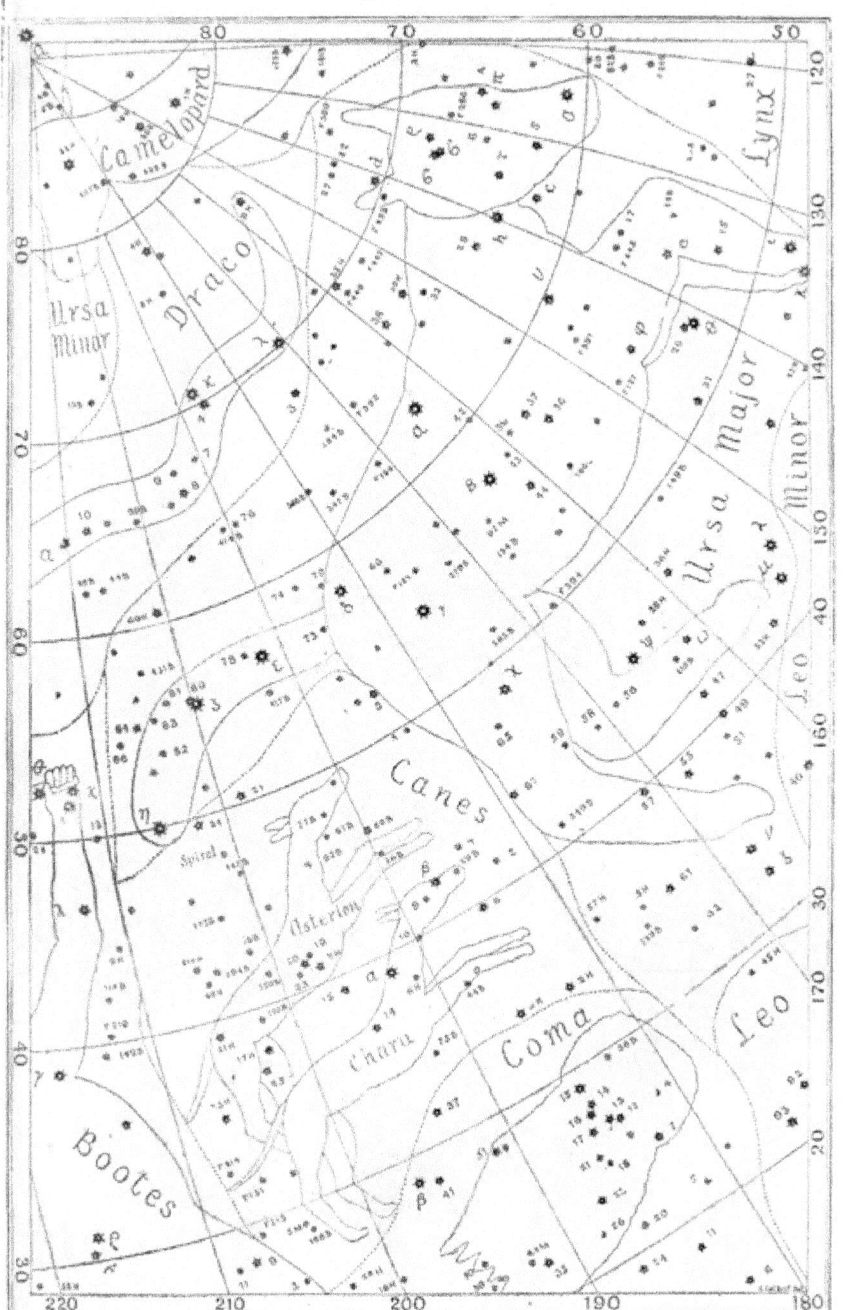

80
70
60
50
120
130
140
150
160
170
180
190
200
210
220
Camelopard
Draco
Ursa
Minor
Lynx
Ursa Major
Minor
Leo
Canes
Spiral
Asterion
Chara
Coma
Leo
Bootes

STAR-STUDIES.

WHAT WE KNOW

OF

THE UNIVERSE

OUTSIDE THE EARTH.

BY E. COLBERT,

EMERITUS ASSISTANT DIRECTOR DEARBORN OBSERVATORY.

CHICAGO:
THE WESTERN NEWS COMPANY.
1871.

ASTRONOMY is, at once, the oldest and the youngest of the sciences. The apparent movements of the luminaries and principal planetary bodies were observed, and their periods approximately known, in the earliest ages of the world's written history; while the vestiges of that primeval lore are deep-grained in the structure of the most ancient languages. Yet very little was known of the true plan or extent of the universe till after the invention of the telescope — 263 years ago — and it is only within the present century that anything tangible has been reasoned out with regard to astronomy, except its mathematics. The greater portion of the chemistry of the science has really been developed within the past decade. So recent are many of the more interesting deductions that a great portion of even the reading world is ignorant of their character and importance.

The following pages contain, in condensed form, the essence of this *modern* astronomy — at a price low enough to place it within the reach of the multitude — and although many facts previously stated are necessarily introduced, many others will be found here which are entirely new. The whole is an amplification of the two last of a course of lectures delivered by me, in February last, at Mattoon, Ill. The first treats of the analysis of light, and the deductions made therefrom in relation to the chemical constitution of other worlds. The second is devoted to a consideration of the biological conditions obtaining on the sun, stars, and planetary bodies; and sketches the past and future history of the universe, as indicated in the conditions of the present. E. C.

STAR CHEMISTRY.

Study of the light ray; velocity of light; wave motion. Polarization; original and reflected light. Measuring the wave. The solar spectrum; color due to wave length; heat, light, and actinism. The fixed lines in the spectrum; their chemical relations; incandescence. Causes of bright and *dark* lines. Elements in the sun; their condition. Relative light, from sun, stars, planets, and nebulæ. Connection between light and heat. Other spectra. Measure, and cause, of solar light and heat. Physical structure of the sun and stars. Sun spots and sun storms. Modern analysis of light.

The most modern branch of astronomical research is star chemistry — a study which belongs almost exclusively to the present generation, though some of the basic principles employed in the investigation date back into the preceding century. Star chemistry indicates to us the conditions under which matter exists in the universe around us; as the mathematics teach us the relations of quantity — of matter or space.

The various problems of celestial mensuration are all dependent upon one simple process. We find the direction which a ray of light makes, as compared with the direction of another ray, or with the line of level at the place of observation. Two rays of light coming from a named celestial object, to two different points on the earth's surface, describe the unmeasured sides of a great triangle, which, when compared with a small triangle, enables us to state the distance of the object. And two rays, coming from opposite sides of the same object, and meeting at the eye, give us the principal sides of a reversed triangle which determines the actual magnitude of the celestial body. Similarly, the direction which a ray of light takes to-day, to meet the eye, compared with the direction of a like ray at any other observed time, furnishes us with all the evidences we possess in relation to the movements of those mighty masses, and the laws which govern those motions.

All that we know of the universe, outside of our earth, and much of our knowledge of things terrestrial, has been obtained by the aid of light. A luminous ray, or a succession of rays, strike upon the eye — an organ furnished us for the express purpose of receiving the sensation of light — and we see; we reason from these sensations, and then we know.

Without any appreciation of the properties of this mysterious agent, which men call light, except its well-known property of moving in straight lines, we have reasoned out all the more important facts pertaining to the distances, bulks, weights, and movements of the heavenly bodies. And there are very many other facts, equally interesting and important, which have been ascertained, during the past few years, through the same agency. We can reason out the conditions under which matter exists in other worlds, and trace their adaptability, or the lack of it, to sustain animated existences similar to those found on our own globe. But in order to comprehend the value of those deductions, it is necessary to know something of the nature of that force which lies at the foundation of all our knowledges, as it formed the basis of our being. The first, and greatest, in the recorded list of creative enactments was, not, let there be matter, but, "Let there be light."

The first great fact about light, is that it moves —occupying a measurable interval of time in passing from place to place. Its velocity is so great that, for small distances, the sensation appears to be received without loss of time; but the interval is appreciable for great distances. Von Römer, a Danish astronomer, first observed that the eclipses of Jupiter's satellites appeared to occur about 16 minutes earlier when the earth and Jupiter are on the same side of the sun, than when on opposite sides of the luminary. The difference in time has since been ascertained to be 16′ 26″.6. This fact can only be accounted for by believing that the rays of light which leave the planet at the instants of apparent contact with his moons, occupy nearly 16½ minutes in traversing the whole diameter of the earth's orbit, 182,865,700 miles. Dividing the distance by the time, we obtain 185,349.4 miles per second, as the velocity of light.

Light is immaterial; not a substance, but a force. A tapping motion in the middle of the surface of a smooth sheet of water will cause a series of waves to form around the point struck, and the waves will spread outward in all directions. If a light object be floating on the water, it will be moved up and down, but not outwards, by the undulation. This shows that the particles of water do not move from centre to circumference with the propagation of the wave; but that the particles set in motion at the centre move up and down, and that the movement is rapidly communicated to adjacent particles, lying outwards from the origin of the movement. The individual particles of water vibrate up and down, only; and a cross section would show a movement at the surface like that exhibited by a rope, one end of which is fixed, and the other moved up and down rapidly. The sensation of light is communicated in this way, by a luminous body, in all directions; and the surrounding air, or ether, is made to vibrate like the water wave.

But in order to represent the real character of

the vibration produced by a ray of sunlight, we must consider it to be made in every direction—simultaneously—to the right and left, as well as up and down. The vibration of the ray is more like the movement of the surface of a muscle, which expands and contracts with the movement of the limb. Suppose we have a round cord of India-rubber, and alternately stretch it, and allow it to contract to its original length. As we extend the length of the cord it becomes thinner, and swells out as it shortens, equally on all sides. If we suppose this cord to represent the length of a single vibration of sunlight, the change in the outline—the changing distance of the surface of the cord from its central line—will give some idea of the vibration of the light ray, in every direction. But when that ray of sunlight has struck upon a reflecting substance, at a certain angle, and is reflected back to the eye, we find it to have lost this universality of pulsation, and to vibrate only in one direction — as up and down, like the water surface. Light so changed is said to be polarized. The same effect is produced when we allow the ray to pass *through* a great number of substances. If we take a thin plate of tourmaline, and hold it up in the path of a single ray of light, coming through a pin-hole into a darkened room, the ray will pass through it freely, and will also pass through a second plate of tourmaline held symmetrically to the first; but if we turn the second plate a quarter round, the ray will not pass through it. The structure of the tourmaline may be represented on a large scale by a picket fence. We could shake a rope up and down between any two adjacent pickets; and the wave motion of the rope would not be interfered with if passed between two fences, one built behind the other. But if the one set of pickets were perpendicular, and the other horizontal, the rope could not be vibrated. The tourmaline stops all the vibrations except those which are in the direction of its own grain; it polarizes the ray. We find that all light coming direct from its origin, whether the sun, or a candle, is unpolarized. An instrument which enables us to tell whether a ray has been polarized, or not, is called a polariscope.

When we examine the light coming from the fixed stars, through the polariscope, we find that it is unpolarized; the fixed stars shine by their own light, like our sun. But the same instrument shows that the moon and planetary bodies shine with a reflected light; the light coming from them is polarized; it has been first received from the sun, and is then reflected to us.

Two or more adjacent light waves coming from the same source, vibrate in harmony. But we can separate them by a simple device, and cause them to travel over paths of unequal lengths; then bring them together again, and thus measure the length of the wave. If we

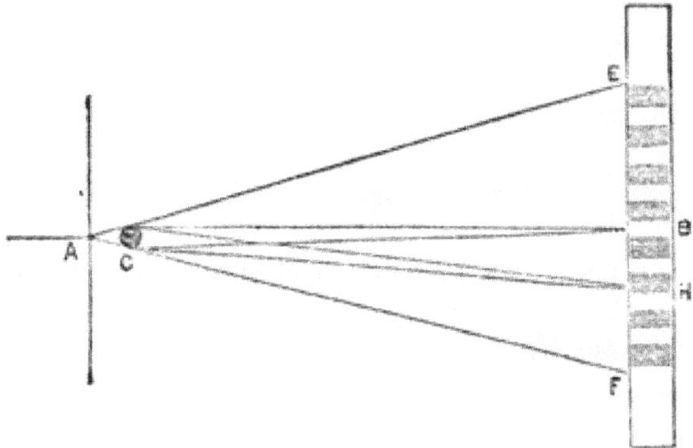

admit a ray of light through a pin hole, A, in a shutter, into a darkened room, and let it fall upon a sheet of paper at E F; then interpose an opaque cylinder at C, it may be thought that the whole space between the lines E and F will be darkened. But we really find that the space is divided up into a number of light and dark bands, the light doubling around the obstacle, as sound can be heard around a corner. The rays which meet in the middle, at B, have traversed lines of equal lengths, and their vibrations are accordant at the point of meeting, producing light; at the point H they have traversed paths of unequal length, so that the swell of one ray meets the reverse part of the other ray, and the interference of the vibrations takes away the effect of light, producing darkness, just as two non-coincident water waves would neutralize each other. At all the other light spaces the rays coming from opposite sides of the obstruction have traversed unequal paths; but the difference of those paths is equal to the length of one or more vibrations, and the rays beat in unison, producing light. If we place an opaque substance so as to intercept the light, either above or below, the bands disappear; showing that *both* sets of rays are necessary to the production of the phenomenon.

The following diagram represents the interference of the rays, C and D, producing darkness; their vibrations destroy each other at the point of meeting. The lines A and B, having a difference equal to the length of one vibration, beat in unison at the point of junction, and produce light. We can calculate the relative lengths of the paths along which these rays have traveled, to agree or interfere with each other, and their differences will give the lengths of the wave pulsations of light. The average length of the wave of yellow light is about 22 of the parts obtained by dividing an inch into one million equal portions.

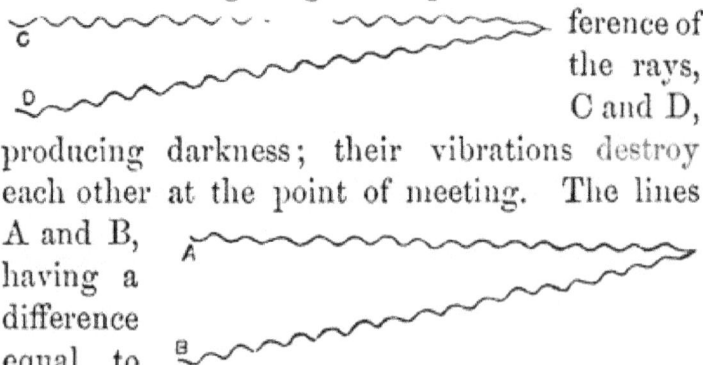

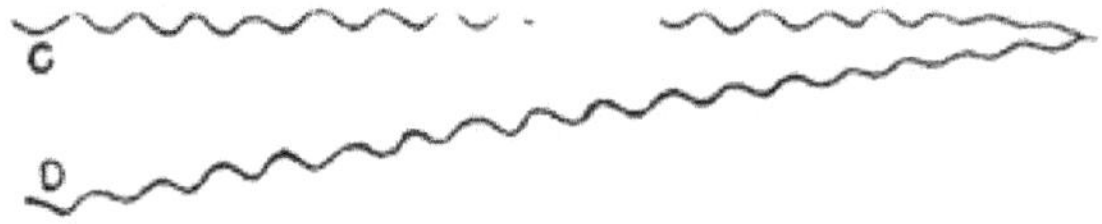

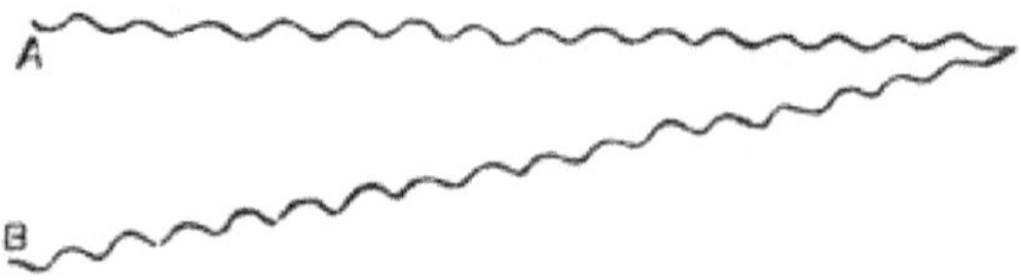

There are about 45,000 pulsations in the length of a single inch. Now, in each second of time light travels over 185,349.4 miles, or 11,743,-737,984 inches; therefore our light ray must make the inconceivable number of 530 million million pulsations in a single second.

The ray of light is not a unit. Sir Isaac Newton discovered that a ray of sunlight may be artificially divided into the seven colors of the rainbow. If we take a triangular bar of solid glass, called a prism, and place that in the path of the ray coming through the small hole in the

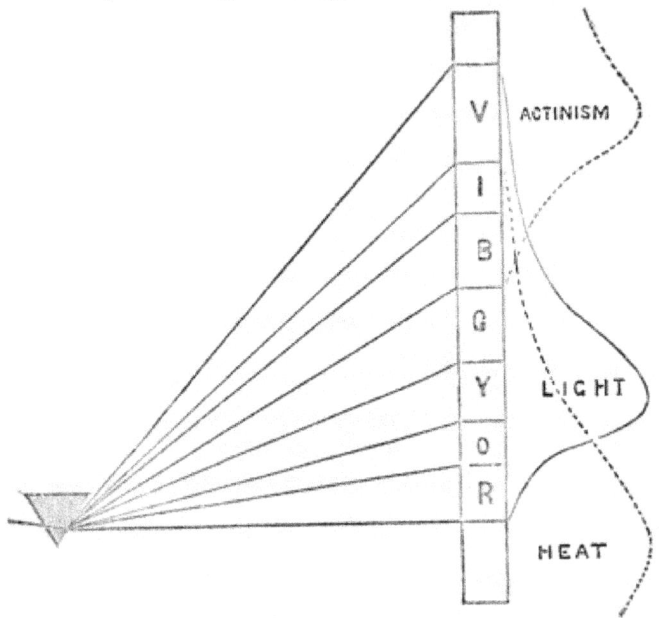

shutter, instead of the opaque cylinder, we find that the ray is split up; and on the sheet of paper we obtain a beautiful elongated rainbow image. Examining it, we find that the several constituents of the solar ray are all bent at different angles from their former course, by passing through the prism. The red ray is least bent from the original course; the violet is most largely deflected, or refracted, from the direction taken by the whole ray before it fell on the prism. The image thus formed is called the Solar Spectrum.

Measuring each of these separated rays, by processes similar to that just referred to, we find that each has its own rate of vibration and wave length, the product of the two quantities being equal in each case. The extreme red is found to have 35,112 wave lengths to the inch; and the extreme violet 64,020 lengths to the inch. The following are the corresponding numbers of vibrations per second, and length of wave, in decimals of an inch, at each end, and at the middle of the spectrum—the junction of the green and the blue.

	Vibrations.	Wave Lengths.
Extreme Violet - -	751,840,000,000,000. - -	.00001562.
Middle - - - - - - -	606,285,000,000,000. - -	.00001937.
Extreme Red - - -	412,350,000,000,000. - -	.00002848.

These are exact measures. The numbers representing the wave lengths for the *middle* of each color are, approximately; Violet, 162; Indigo, 173; Blue, 186; Green, 202; Yellow, 220; Orange, 242; Red, 270; ten-millionths of an inch.

The different colors of light are, therefore, considered to be due to the different lengths of the wave.

The colors are not equally distributed. If we divide the visible spectrum into 360 equal parts we find the following to be the numbers representing the lengths of the different colors: Red, 45; Orange, 27; Yellow, 48; Green, 60; Blue, 60; Indigo, 40; Violet, 80. We may reverse the analysis by a very simple process, and show that the blending of the rainbow colors in these proportions produces white light. On the flat side of the rim of a wheel paint the seven colors in spaces proportional to the above numbers; if you turn the wheel round so rapidly as to prevent the colors from being seen separately, the painted rim will appear to be a pure white.

It is a singular fact that, though the numbers we have given represent the limits of the visible spectrum, yet numerous experiments prove that the diffused rays extend some distance beyond the red, and reach out past the violet over a space many times greater than that occupied by the visible spectrum. We can see the spectrum for some distance past the violet, by allowing the rays to fall on certain preparations made for that purpose. The visible spectrum is really the light-giving portion of the whole, which is found to exhibit three distinct, and partially separate properties. If we hold a delicate thermometer in the path of the red rays, the instrument shows a rise of temperature, and the invisible rays beyond the red are more potential in the production of heat than those which are visible; while scarcely any heat is manifested in the violet region. The solar ray is also a cause of chemical change; a fact which is taken advantage of by the photographer in preparing his pictures. If we place a piece of paper, spread over with chloride of silver, in the spectrum, it is soon blackened, but not with equal rapidity in all portions. It changes color speedily in the violet region, and beyond it; but changes very slowly if held among the red rays. The greatest amount of light is found near the middle of the spectrum. We can readily obtain from the solar ray two out of the three forces, heat, light, and chemical activity, isolated from the others. We can obtain heat without light by placing the thermometer beyond the red end of the spectrum; chemical action without light or heat, by placing a sensitized plate beyond the violet; and light, almost without the other two, nearly at the junction of the blue and the green. We can not eliminate any portion of the effective properties of any part of the ray, but we can isolate one pulsation, or group of pulsations, from all the rest.

The curved lines on the right hand side of the

last diagram represent the relative intensities of these three forces in different parts of the visible spectrum; the force being proportional to the distance of any point in the spectrum from the corresponding points in the curves. Thus; the average light intensities for the several rays are, Red, 94; Orange, 640; Yellow, 1,000; Green, 480; Blue, 170; Indigo, 31; Violet, 6.

It was for a long time thought that the seven colors are, themselves, simple; the well-known fading of one into the other, in the spectrum or the rainbow, being accounted for by the supposition that the colors lap over each other near the places of junction. It has also been taught that there are only three primary colors — the red, the yellow, and the blue — all other colors and shades being produced by the admixture of these in various proportions. But the researches of modern science do not permit us to accept either of these theories. We find that the wave length varies gradually from one end of the spectrum to the other; and conclude that there are really thousands of waves concerned in the production of the spectrum, each giving a color of itself, distinct from all the rest, and having a wave-length measurably different from all its fellows. The blending of all produces white light; the absence of all is darkness; the union of many contiguous waves gives one of the seven prismatic colors. The ray of solar light may be regarded as a bundle of independent, yet connected immaterial pulsations, each of which has its own wave-length, by virtue of which it possesses individual powers for the production of the sensations of heat, color, or chemical action—named in the order of increasing refrangibility, and decreasing wave length. In passing through the prism these pulsations are separated, and made to diverge, so that they can be viewed and experimented on independently. We can not say "viewed separately," for it is not improbable that the single ray is composed of a greater number of pulsations than can be counted—just as the smallest living form is made up of millions of atoms of matter.

If, instead of a round hole, we make a very narrow slit through which to admit our ray of sunlight, we find the dispersed image to be crossed, at irregular intervals, by fine, shadowy lines. Coming through the circular opening the dissected light is thrown on the spectrum in a series of circular patches which overlap each other, much in the same manner as a spread-out pack of cards will do, and thus produces a partial blending of effects. The phenomenon of overlapping is avoided when the slit is used. Dr. Wollaston first noticed this fact, distinctly, about the beginning of the present century; and a German optician—named Fraunhofer—mapped out the positions of 576 of these lines about the year 1814, and designated some of the most prominent by the first eight letters of the alphabet. The number of lines in the solar spectrum is about 2000. Fraunhofer afterwards examined the light coming from several of the fixed stars, and found different arrangements of lines from those noted in the sunlight; this was accepted as a conclusive answer to the charge that the dark lines are caused by the absorptive properties of the earth's atmosphere; it was concluded that the cause of these diversities could only be found in the bodies examined — far outside the ærial envelope of our globe. This was really the first step in the spectrum analysis, which has since become so important a branch of research, and has opened up to the eyes of mortals the grandest of truths, while it has placed in their hands a divining-rod with which to explore the most hidden mysteries of earth and heaven.

In order to understand the value of this discovery it will be necessary to look at the chemical constitution of things on this earth. The material world appears to us under an endless variety of forms, but the skill of the chemist has shown that all are built up from a comparatively small number (67) of substances, called elements. We are acquainted with some of these elements, as iron, gold, silver, copper, and zinc, but the majority exist always in combination with others, and are only isolated artificially. Thus: we do not find oxygen alone; but we meet with it mixed with nitrogen, to form the air we breathe; combined with hydrogen to form the water we drink; with aluminum to form clay; with carbon in the breath we exhale from the lungs; with hydrogen and carbon to form sugar, and pure alcohol. Salt is not a simple substance, but is composed of two elements, named sodium and chlorine. Now, it is a singular fact that every one of these elements, when made to emit light, under certain conditions, gives a peculiar arrangement of lines in the spectrum, which is not given by any other element, and is given by itself, whether burned singly or in combination, however intricate. Whether we subject a drop of human blood, or a lump of salt, or a metal bathed in sea-spray, or any other of the shapes in which sodium is found, to the spectrum analysis, we equally find a double line, nearly in the middle of the yellow portion, which indicates the presence of sodium, however much it may be disguised to our senses, and though existing in the smallest possible proportions in the compound; the double yellow line in the spectrum reveals its presence. The spectrum analysis has discovered to the chemist the existence of no less than five new elements—all previously unknown.

The world principally owes to the labors of

Bunsen and Kirchoff, the development of the processes by which these facts have been usefully applied to the study of astronomical physics; though Angstrom, Roscoe, and Zöllner, have also contributed largely to our fund of knowledge on this topic. They discovered that a burning solid gives a continuous spectrum, unmarked by lines; but that when the solid is reduced to the gaseous condition, the burning gas gives a spectrum (except under certain well understood circumstances) which is crossed by the bright lines characteristic of the substance. They next studied the image produced by the light of the burning solid and the burning gas combined; the latter being placed nearest to the prism. They arrived at the remarkable conclusion that the characteristic rays of the substance, as given out by the solid, are neutralized by passing through the burning gas, which seems to be transparent to all but the characteristic lines given out by itself. The lines in the spectrum were faithfully reproduced; but they were *dark* — just like those visible in the spectra of the sun and stars.

The instrument used in these investigations is called a spectroscope.

It is found that most of the elements give more than one line — some of them very many — but whether few or many, the positions are always the same for the same elements. If a single element be subjected to the test, the lines of that element are visible; if a compound of two or more elements be examined, we find the lines which belong to each of them. Hence, as a rule, the more complicated the substance analyzed, the greater is the number of lines visible at one time in the spectrum. Whether in combination, or apart, each element, when submitted to the spectrum test, gives indubitable proof of its presence, by the production of lines which are not given by other elements. It is, however, true that some of the elements interfere more or less with the lines of others, when in combination; the blue line of strontium disappears from the spectrum, if the chloride of that element be burned with the double chloride of copper and ammonium. We are not, therefore, justified in assuming that no elements are present in a burning body, other than those which give their fixed lines in the spectrum; while we *are* warranted in believing that all those are present which show their characteristic bars on the ribbon of dissected light.

The accompanying diagram shows the principal lines in this solar spectrum; it would not be possible to represent the whole of the 2,000 lines, or more, on so small a scale. The lines C, F, and H, are those which show the presence of hydrogen. The line D, is double, falls in the

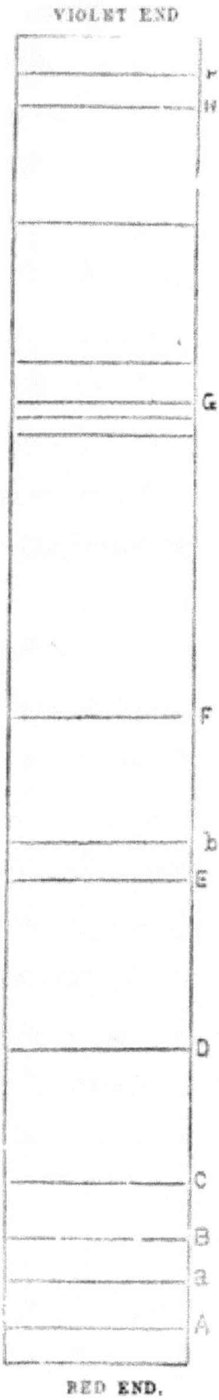

VIOLET END

middle of the yellow, and is the well-known sodium line. Iron makes itself known by a group of seven lines, of which E is the principal one, by a smaller group just below F, and two lines at G. Magnesium is indicated by three lines at b; and Nitrogen by two lines nearly half way between b and F.

The solar spectrum contains the lines which indicate that the following elements exist in the Sun, in a state of incandescence: Sodium, iron, hydrogen, calcium, barium, chromium, nickel, copper, magnesium, cobalt, zinc, cadmium, manganese, aluminum, strontium, titanium, gold, and potassium (18). Of these, cobalt, strontium, cadmium, and potassium are regarded as not clearly shown to exist there. Other metals may exist in the Sun, but, if so, in smaller proportions in the atmosphere, or in larger quantities but not vaporized. Hydrogen, which is the lightest of our known gases, and is now believed to be a metal at extremely low temperatures, is present in very large quantities.

RED END.

No trace of oxygen has yet been discovered in

the sunlight: if that element exists in the Sun it must be in very much smaller quantities than on the earth, as if present in considerable proportion it would undoubtedly exist in the gaseous surrounding, as in the case of our earth.

From these facts we are enabled to draw the following important conclusions:

The Sun is composed of the same materials as those which form a large proportion of the matter which makes up the bulk of our earth, and of all that exists on its surface; while, yet, some of the most important of our earth elements appear to be absent.

The Sun is intensely hot. The heat at his surface must be at least as great as that which attends rapid combustion here; because the fixed lines in the spectrum appear only when the body examined is at a very high temperature.

The heat at the solar surface is great enough to reduce to a state of vapor many of the elements which are fusible only with great difficulty, by the aid of the greatest heat we can produce artificially. The vapor of iron which is detected in the Sun's atmosphere, necessitates a temperature of not less than 4500° Fahrenheit.

The Sun is a globe of incandescent solid or fluid matter, surrounded by an incandescent atmosphere. The lines in the solar spectrum would not be dark, if it were otherwise. In this interior mass — the Sun proper — a great number of chemical elements may exist which do not yield up vapor to the surrounding atmosphere, just as there are elements in the earth's structure which are not found in our air; but those elements which are gaseous with us, would most probably be apparent in the solar spectrum if they existed in the Sun. All the elements thus far traced in the Sun exist within as incandescent solids, or fluids, and without as incandescent gases. We shall see presently that the Sun's interior is necessarily fluid — rendered so by intense heat.

The spectroscope enables us to ascertain that this incandescent atmosphere extends to a distance of five to six thousand miles from the Sun's surface. This shell of incandescent gas is called the chromosphere.

[In an article in the March, 1871, number of the "Lakeside Monthly," I wrote as follows:]

"Outside of this another phenomenon is visible during a total eclipse of the sun. It is the corona (crown), which resembles the halo some of our painters have depicted round the heads of the saints. The shape and extent of the corona are not the same in all total eclipses of the sun; in August, 1869, it was nearly twice as broad as in December, 1870. The limit of breadth of the true corona appears to be from one-fifth to two-fifths of the sun's apparent diameter. The solar diameter being nearly 853,000 miles, two-fifths of this quantity is about 340,000 miles, which is, approximately, the greatest breadth of the true corona — that which gives in the photograph an evidence of its existence.

" Very recently it has been surmised that the coronal display is caused by the reflection of the sunlight from numberless planetoidal chunks or specks of matter, which are either revolving around the central luminary in very small orbits, or else are falling to his surface to keep up the supply of light and heat at the centre of our system of worlds. This proposition is absurd. By means of the third law of Kepler — that the squares of the times are proportional to the cubes of the distances — we can easily compute that a mass of matter 350,000 miles distant from the sun's surface, must revolve around him once in a little less than seven hours (6.87) to preserve its orbit. That is, the mass of matter in question would swing round the sun nearly ninety times while the sun turns once on his own axis. Now, all the analogies of the system, and the very laws of motion, as we understand them, teach that a body revolving around the sun must have a *less* angular velocity than the sun; and that if the angular velocity be the same, it is *already* a part of the sun. The corona can not, therefore be produced by reflection from bodies, whether large or small, moving in an orbit. We need scarcely to refer to the other branch of the theory. To suppose that matter which once circled around him, is continually falling to the sun in such immense quantities as would be necessary to produce this phenomenon whenever an eclipse permitted us to see it, would be to suppose that the solar system is in a state to which a galloping consumption is but a snail's march; it is simply impossible.

"Our atmosphere is usually spoken of as lying within forty-five or fifty miles of the earth's surface. Now, observations made of the height of the aurora polaris of August 28, 1859, at several stations situated on a line extending from the West Indies to Maine, showed that the aurora formed a stratum of light extending from forty-six miles to five hundred and thirty miles above the surface. The aurora is now known to be due to an illumination of atmospheric particles by electric excitation, just as the rainbow is produced by the sunlight reflected from the rain-drops. The auroral display is believed to be seldom visible at a less elevation than forty-five miles. In other words, it is only produced by the operation of the electric force on matter so extremely tenuous that a globe of it, the size of our earth, would not weigh a single ounce. And it is easy to calculate that the atmosphere extends much further than the visible limits of the aurora. The rotary motion of the earth on her axis gives a centrifugal motion

2

which destroys one part in two hundred and eighty-nine of the weight of a body at the Equator, and we must rise to a height of twenty-two thousand miles above the surface before we come to a point where the attraction of gravitation to the earth would be completely balanced by the centrifugal motion. Inside of this limit, a body must pass round the earth more rapidly than once in twenty-four hours to avoid falling to her surface. Outside of those limits, a body must move more slowly than the earth's axial velocity in order to retain its distance. That distance is, therefore, the limit of the earth as an integral mass. All within it necessarily belongs to the earth, and the process of expansion must proceed from the surface to that point, before we leave the attenuated atmosphere and find ourselves fairly launched into the regions of space. The air is sufficiently dense to be susceptible of illumination, only to a height corresponding to one part in forty-two of this distance.

"Computing the sphere of the sun's attraction on the same principle, we find that it extends a little more than fifteen millions of miles from his surface. The same ratio as obtained in the case of the earth (1–42) will give 368,000 miles from the sun's surface for the limit of the true corona, or 13¾ minutes of arc at the sun's mean distance. My sketch of the corona during the eclipse of 1869, made at Des Moines, Iowa, shows the corona to have extended to about this distance all around the sun. It seemed to extend a little further than this in one or two directions, in pointed rays of hazy light; that was undoubtedly the result of earth atmospheric conditions.

"We have here a precise analogy between the conditions of the earth and sun, which warrants us in concluding that the corona is a genuine solar phenomenon; and that, whether the portion nearest to the disc be self-luminous or not, the exterior portion, to a distance of not more than 368,000 miles, is produced by the reflection of solar light from a solar atmosphere.

"But there is one other point of similarity between the two sets of conditions in earth and sun, which tends to establish the conclusion I have essayed to deduce. It is that the corona gives a line in the spectrum which corresponds with the division of 1474 in Kirchoff's scale, and is not given by the burning of any chemical element with which we are acquainted, but is seen in the spectrum of our aurora polaris. In both cases this indicates the existence, in the higher regions of the atmosphere, of an element which is probably distinct from all those known to us, and takes the more elevated position by virtue of its less specific gravity; as hydrogen takes an exterior place in the sun for the same reason, and as it possibly does in our atmosphere, though the fact has not yet been accepted. We have no warrant for supposing that this, or any other line, gives us a trace of that mysterious ether which fills all space not occupied by denser matter; that ether is too much attenuated, and too cold by reason of that attenuation, to give the semblance of a line in the spectrum.

"A very interesting observation was made by Professor Young during the eclipse of Dec. 22, 1870. Directing his spectroscope to the eastern edge of the sun, just as the advancing moon touched it at the beginning of the total phase, the dark lines in the spectrum suddenly died out, leaving the continuous spectrum, which had never been seen previously except by Secchi, the eminent Italian astronomer. This condition lasted but for an instant, and then the lines which were dark in the ordinary spectrum suddenly became *bright*, and continued so for one or two seconds of time. Now this latter fact not only proves that the theory of Kirchoff is true — that the *dark* lines are due to the incandescence of a solid or liquid, on which is superposed an incandescent gas — but it also leads to the valuable deduction that the space of about four hundred miles in depth at the base of the chromosphere is the only one which *contains* all the elements the existence of which in the sun is revealed by the spectrum analysis.

"I conclude, therefore, that the facts in regard to the constitution of the sun are as follows:

"First. The sun consists of an incandescent liquid mass, corresponding to the liquid interior of our own globe. That mass — the dimensions of which are usually accepted to be those of the sun itself — contains eighteen chemical elements known to us, and probably many more, the existence of which is not shown, because they do not also exist in the gaseous condition outside. Our earth's interior *may* contain several elements which have never been met with on the surface. If the solar system originally formed one vast mass, which was afterwards separated into numerous bodies, it is but reasonable to suppose that the central orb contains most of the elements found in the planets, and some which are peculiar to himself.

"Second. This more dense fluid mass is surrounded, to a depth of some four hundred miles, by an incandescent gaseous envelope which corresponds in position to our earth's crust. This envelope contains about eighteen of the chemical elements which exist within; and its superposition on the central mass is the cause of the dark lines in the solar spectrum.

"Third. This shell is surrounded, to a depth of five or six thousand miles, by the true chromosphere, which corresponds to our ocean, and contains but five or six of the elements, of which hydrogen appears to be most abundant — all in an incandescent state.

"Fourth. A non-luminous envelope, corresponding in position *and function* to our atmosphere, which is dense enough, for a height of nearly four hundred thousand miles, to reflect the light emitted from the incandescent interior. This reflected light is too feeble to be appreciable by us in competition with the glare of the direct sunlight; but is visible when that direct sunlight is cut off by the interposition of the moon. Vast masses of hydrogen, and the other components of the chromosphere, shoot up into this envelope, as the waves of ocean dash up into our atmosphere, and are carried above the normal surface in a way which finds a feeble analogy in the cloud formations of our atmosphere from the waters beneath. The light reflected from this envelope all gives evidence of polarization, except that coming from the portion immediately adjacent to the chromosphere. The latter undoubtedly both emits and reflects light.

"Fifth. The extent of the coronal display, during a total eclipse of the sun, corresponds to the relative intensity of the convulsions occurring nearer the normal surface, which are gauged to us by the comparative extent of the sun spots, and the magnitude of the rose-colored protuberances visible during the total eclipse, and the faculae which are seen at other times in the telescope. As these latter phenomena of solar variability are now identified with the positions presented by planetary revolution, the extent of the corona is referable to the action of worlds outside the sun, just as the intensity of our auroral exhibitions is also known to be associated with, and due to, the same agencies. The solar corona and the Telluric aurora are, therefore, not only as nearly identical in character as is possible under such widely different sets of conditions; but both vary in the same ratio as their same exciting causes."

It is very seldom that we can bear to look at the Sun with the naked eye. Only when his rays have to pass through considerable quantities of cloud vapor before reaching us, can we view him without the aid of smoked or colored glasses, to modify the effects of his blinding glare. The most intense lights we can produce artificially, are nothing as compared with it. The most brilliant artificial light, the calcium, produced by a ball of quicklime, on which a mixture of ignited oxygen and hydrogen gases is kept constantly playing, can not be looked at without injury to the eye, if brought near enough to appear of the same size as the Sun. But if the calcium light be placed between the eye and the Sun, and both be observed through a colored glass, the calcium light looks black by comparison. Sir John Herschel estimated that the Sun gives out as much light as 146 balls of calcium, each the size of the Sun, would give, if removed to the same distance from us.

Let us take as the unit of light measurement, the faintest star which is distinctly visible to the naked eye on a clear night. There are about 4424 such stars, classed as of the 6th magnitude. There are 959 stars, each of which gives twice as much light as this unit, and are classed as of the 5th magnitude. In the order of 4th magnitudes we rank 327 stars, each giving 6 units of light. In the 3rd magnitude there are 141 stars, each giving 12 units of light. Of the 2nd magnitude there are 34, each giving 25 units of light; and of the 1st magnitude there are 20, each of which gives 100 or more units of light. Strictly speaking, there are very few stars of equal brightness, but it is found convenient to classify them as nearly as possible to these averages. Sirius gives 320 such units of light, and Alpha Centauri, the brightest star in the Southern sky, except Canopus, and the one which is believed to be nearest to our Sun, gives eighty times as much light as an average star of the 6th magnitude. The light given to us by the planetary bodies varies widely with their distances from us. We may represent the averages by the following numbers: Uranus, 1; Saturn, 25; Jupiter, 250 (sometimes 500); Mars, 30; Venus, 160 (sometimes 600); and Mercury, 6. Now, if we add all these together, and allow 4000 units for the filmy light received from the milky way, and other stars too faint to be individually visible, we shall only have about 20,000 units, or 200 times as much light as is given us by an average star of the 1st magnitude.

The light of the full Moon is estimated to be 9,400 times greater than that received from Sirius, or 3,000,000 times our unit, or 150 times as much as we should receive from all the planets and stars together, if they were all above the horizon at one time. But the light of the Sun is computed to be 547,500 times greater than that of the full moon, or equal to that of (1,642,500,000,000) more than one and a half million million stars of the 6th magnitude, or 8,212,500 times greater than that of all the planets and stars together. We receive sixteen million times as much light on a clear day, as on a star-lit night when there is no moon.

And yet we have good reason to believe that very many of those shining bodies, which contribute such infinitesimal proportions to the light and heat we receive on this Earth, are much larger than our Sun, and radiate many times more heat into space than does the centre of our system. The light of the Sun is about ten thousand million times greater than that of Sirius, but if the star were brought as near to us as is the Sun, we should receive nearly eighteen hundred and ninety thousand million times more light from him than now; and we find in this way that Sirius gives off one hundred and

*

ninety-two times as much light as our Sun; whence his diameter must be fourteen times greater, and his volume 2688 times larger, if the light of both be equally intense for each square yard of surface.

Contrast with this the light of the nebulous clusters which are visible only in the telescope. Their light has been estimated to range from $\frac{1}{1500}$ to $\frac{1}{200000}$ of that of a sperm candle at the distance of one quarter of a mile from the eye.

We have already seen that light travels at the rate of 185,349.4 miles per second. Yet with this velocity the light which an inhabitant of the Southern hemisphere receives on his eye from Alpha Centauri, left that star nearly 3½ years previously. It is estimated that, on an average, light requires 15½ years to reach us from a star of the 1st magnitude; 28 years from a star of the 2nd magnitude; 43 years from a star of the 3rd magnitude; 120 years from a star of the 6th magnitude, and 60,000 years from some of the faint nebulæ to which we have just now referred. Though only a faint glimmer of light is able to struggle into view from these shadowy masses, in the telescope, yet these nebulous clusters of matter must be immense to be visible at such great distances. The nebula in the sword of Orion can scarcely be less than 100 times the diameter of the whole orbit of Neptune; and the nebula in Andromeda has a diameter fully three times greater than that of Orion. The nebula of Andromeda is at least eight hundred thousand million miles in diameter, and a beam of light would occupy more than two months in passing through it; the dimensions of the nebula *may be* even 400 times that immense number.

Light and heat are co-ordinate; but they do not always correspond in intensity, as we have already seen in looking at the several divisions of the solar spectrum. If we compare the results of combustion in the case of hydrogen and carbon, we find that hydrogen gives out the most heat of the two, for equal weights burned. So with the stars; which, though not necessarily burning, in the sense in which substances burn on this earth, are yet all in a fervid glow,—heated so highly that nothing with which we are acquainted could retain the solid form if transported to one of those incandescent masses. The great majority of the bodies we see are intensely hot; too hot to permit vital existence, *as we understand it*, upon their surfaces. But we must remember that at the distance of Sirius, the Earth, and other planets of our system, would be invisible. Each of those stars, like our Sun, may be a centre of light and heat to a family of worlds, though we see them not.

A very good idea of the connection between light and heat may be obtained from viewing the successive phenomena produced by the gradual heating of a bar of metal in the fire. A piece of iron, heated to the point of boiling water, emits rays which can be felt by the hand, though invisible to the eye. At the temperature of 977° the rays become visible — the iron appears red; and if we place it before the prism the red rays appear in the spectrum, unaccompanied by other colors. As we increase the temperature from this point the color of the iron changes, and the different hues appear consecutively in the spectrum; successively, counting from the red end, till at the temperature of 2130° all the colors appear at the same time, and the iron becomes white hot, the whiteness being due to the blending of all the rays, which are individualized in the spectrum. At still higher temperatures the color is white, showing that all the prismatic hues are given out at any temperature over 2130°, at the same time; but the chemical activity of the rays beyond the visible violet is heightened, and increases with the heat imparted to the iron, which must be raised to a very much higher temperature in order to prolong the chemical rays to the same distance as in the Solar spectrum. The characteristic lines of the iron appear in the spectrum, successively, in the same order. We arrive at the singular conclusion that the greatest chemical activity, and the darker colors, are only produced by the most intense heat; and that if the

Sun were only as hot as melted iron there would be but very little chemical change possible on the Earth's surface — no growth of vegetables or flowers, or flesh and blood — no vegetable or animal decay. The heat rays are really those given out at comparatively low temperatures.

In this fact we have a means of forming some idea of the comparative temperatures of the stars. Those which give out rays in which the red predominates, as Aldebaran, Betelgueuse, and Antares, are perhaps relatively cool bodies; and the white stars, as Polaris, Regulus, Denebola, and Fomalhaut, are among those which give out the most light, are the hottest, and their attending planets are those among which the most rapid changes take place, owing to the greater amount of chemical activity developed.

We may dwell a little longer upon the spectrum, and find that it has still many secrets to reveal. We turn our instrument to the stars, and we find that they all, like our Sun, contain elements which are found on the earth; but that the exhibited elements are not the same in all. In the spectrum of Betelgueuse, we have lines which show the presence of sodium, magnesium, calcium, iron, and bismuth, in the star. In the light of Sirius we find the lines which tell of sodium, magnesium, iron, and hydrogen. In Pollux and Vega we have evidences of the existence of sodium, magnesium, and iron. Aldebaran has hydrogen, sodium, magnesium, calcium, iron, bismuth, antimony, and mercury, with the element called tellurium, which was so named from Tellus (the earth) on the supposition that it is peculiar to our globe. There is no tellurium or bismuth in the Sun. Hydrogen, which is present in such vast quantities in the Sun, and on the Earth, is not found in Betelgueuse.

The spectroscope also enables us to find reasons for the different colors of stars, in the special arrangements of lines in their spectra. Thus: Our Sun, Pollux, Capella, and Tarazed, are yellow; giving but few dark lines in the yellow portion of the spectrum, to interfere with the full effect of the yellow ray. Ras Algethi (Alpha Herculis) is an orange-colored star, because it has few shadowy lines in the yellow and orange portions of the spectrum; the great majority of the lines falling in the green, the blue, and the red.

If we direct the spectroscope to the nebulæ, we find some startling differences. Some of them exhibit a spectrum analogous to those of the fixed stars, showing that they are composed of aggregations of luminous bodies too far distant to be individually visible. But we find that the spectra of many nebulæ are made up of but one color, crossed by two or three bright lines. We have already seen that the *bright* line is produced by a glowing gas, without an incandescent solid or liquid mass behind it, or in its interior; and in this peculiar spectrum we find not only a proof that these nebulæ are only aggregations of gas, or cloudy matter, but a fact which overthrows a theory till recently thought to be well established. It had been found that the higher the power of the telescope employed in looking at these shadowy masses, the greater was the number of those resolvable into small stars; and it was inferred from this, that all those nebulæ which appeared only as films in the field of the best telescope, were simply assemblages of stars too remote to be seen distinctly. But we now know that many of these masses are not, and never were, stars; though they may become such in the far distant future: in the same way as our solar system is believed to have been condensed from a globular mass which once filled the whole space now bounded by the orbit of Neptune, and had then a density 200,000,000 times less than that of hydrogen. The lines in the spectra of these nebulæ lead us to infer that they are largely composed of hydrogen and nitrogen. It has also been found by recent observations that some of these true nebulæ change their apparent places much more rapidly than the fixed stars in their immediate neighborhood. This fact indicates that the nebulæ are much nearer to us than the stars. Indeed, it has recently been asserted that many portions of the milky way are much nearer to us than the stars which obscure the brightness of that brilliant belt in their apparent vicinity.

Looking at the comets through the spectroscope, we find that they, too, are masses of vapor; a fact apparent, also, in their comparative absence of weight when balanced against the planets and their moons. Incandescent carbon, in more than one form of chemical combination, is believed to be the principal element in many of the comets. The spectroscope fails to inform us with equal distinctness of the composition of the Moon and planetary bodies, because the light we receive from them is only reflected sunlight. But a comparison of the same sunlight, as reflected from different bodies, enables us to draw many valuable inferences respecting the character of their atmospheres, and to see that the Moon has no atmosphere — there is neither water nor air on the Moon's surface.

We have seen that the Sun and stars are intensely hot. Have we any means of answering the question, how hot is the Sun? and thence of forming an idea of the temperatures of the stars?

The experiments of Pouillet (a French philosopher) with the lens pyrheliometer, show that the heat annually received from the Sun by our Earth is equal to that required to melt a layer of ice 101 feet in thickness; and that the heat received from space would melt another layer of ice 82 feet in thickness; only a very small proportion of the latter coming from the stars. The heat annually received, therefore, averages that required to raise the temperature of 183 feet of ice through 140°. This melting power is evidently expended on every point of the surface of a hollow globe surrounding the Sun at the same distance as the Earth; the total area of such a globe would be 2,127,000,000 times the area of a great circle of the Earth. The Sun gives off 2,127,000,000 times as much heat as would be required to melt an ice envelope, 101 feet in thickness, all around the globe. In other words, each square foot of the Earth's surface receives from the Sun, every hour, enough heat to raise the temperature of one pound of water 100°. Now, if we should reason from this that the heat increases as the square of the distance from the Sun decreases, then, the distance from the Sun's centre to his circumference being only $\frac{1}{214.4}$ that of the distance to the Earth, the square of this number, nearly 46,000, would be the ratio of heat given off from every square foot at the Sun's surface; equal to 4,598,940° per hour, or 76,649° per minute. The heat at the Sun's surface would raise one pound of water 1277.5° in temperature each second, on each square foot of the Sun's surface. The ratio being 46,000; then, if the Earth's average temperature be taken at 58°, it might be thought that the Sun's temperature is 58 × 46,000, or 2,668,000°.

By reasonings similar to this, Pouillet has estimated that the heat at the surface of the Sun is sufficient to melt daily a layer of ice $10\frac{1}{2}$ miles in thickness; Sir John Herschel estimates that the heat emitted hourly is equal to that which would be produced by the combustion of six tons of coal on every square yard of his surface, and assumes that the heat is sufficient to keep melting a cylindrical pillar of ice 45 miles in diameter, fed into the Sun with the velocity of light. Dr. J. B. Mayer has argued that the temperature of the Sun ought to decrease by radiation to the extent of $3\frac{1}{4}°$ annually.

But these assumptions are all based on the supposition that the heat received on our globe is equal to the amount received in space at the same distance from the Sun; whereas we know that the amount of the sensation of heat depends upon the medium through which it acts. The Sun's rays always warm us the most powerfully when the atmosphere is most dense; and this is evident, not only from an appeal to the senses, but from a consideration of the character of light. The more dense the air, the greater is the number of air particles which are set in motion in a given space, by the pulsations of the light-wave. Hence it is that snow is never absent from the tops of high mountains, even under the zenith Sun of the equatorial regions, though nearer to the Sun than the luxuriant valleys at their bases. A series of balloon ascensions, made some time ago in England, showed that up to the height of one mile the temperature decreased much more rapidly than the barometric pressure; which fact is doubtless attributable to radiation of heat from the surface. But above the height of one mile the temperature decreases steadily in the exact ratio of the air's density, which is known to decrease with an augmentation in the altitude. We are warranted in concluding that where there is no matter there is no heat. It is an established fact that heat is the result of the vibration of material particles, and where there are no particles to be vibrated there can be no manifestation of heat. Hence the temperature of the interplanetary spaces must be very nearly that of absolute cold, because they contain but very little diffused matter. We can demonstrate that outside of the cometary masses, the ærolitic particles, and the planetary bodies, with their atmospheres, the whole globe bounded by the orbit of Neptune, does not contain so much matter as would weigh a single grain.

There is only one way in which we can hope to measure the temperature of the Sun; the way was first pointed out by me in June, 1870, but it is far from giving precise results, as yet, owing to our want of knowledge of the absolute zero of heat. The zero of our present thermometric scale is entirely arbitrary, being about one-sixth as far below the freezing point of water, as the distance from the freezing to the boiling point. Sir John Herschel estimated the temperature of the planetary spaces to be not less than 260° below the freezing point, and Pouillet places it 256° below, while some recent experiments on the gases indicate about 460° below the freezing of water, as the point at which they lose their elasticity. Assuming 260° as the true quantity, and that the Earth's average surface temperature would be 40° below the freezing point but for the warming influences of the Sun, we have the difference of 220° of the Fahrenheit scale, as the heat of the Earth's surface above absolute zero; independent of the average of 66° received from the Sun to bring her average temperature up to 58° of the thermometer.

Now: what is it that causes the Earth's independent surface temperature to stand 220° higher than that of the interplanetary spaces? We have already seen that the temperature of the air varies with its density, and the same is true of the Earth. The sensation of heat is due to the mutual pressure of contiguous atoms. Hence, when those atoms are widely sundered, there is but little heat; when they are forced together more closely, the sensation of heat is produced,—as may easily be proven by compressing any given quantity of air into a less space than it formerly occupied. We have good reason to know that all matter is in a state of continuous vibration, like the motes which we can all see dancing in the sunbeam, even in a still atmosphere. When we increase the heat of a body we do it by increasing the tendency of its constituent atoms to move; or, in other words, by causing them to enlarge their arcs of vibration. This increases the pressure of the atoms upon each other, and this, again, results in the well-known augmentation of bulk which accompanies an increase of temperature, causing liquids to boil over, and the mercury to rise in the tube of the thermometer. All matter increases in bulk as it becomes warmer; and we can always measure the amount of heat by measuring the bulk of the warmed or cooled body, and comparing it with the bulk at a known temperature, though all bodies do not expand equally with equal additions of heat. Solid bodies expand as the temperature rises, till they reach a point where the repelling power becomes greater than the force which binds the atoms together in solid form, and they become fluid; water takes up 140° of heat in passing from the solid to the liquid form. Liquids, too, expand with augmentation of temperature, till the mutual repulsion of their atoms becomes so great as to overcome the attraction of the liquid form, and then the body becomes a gas or vapor. One cubic inch of water expands into one cubic foot of steam, and takes up 972° of heat in making the change; the constituent atoms are driven so much farther apart that their arcs of vibration can be increased by the amount represented by 972° of heat excitement, before the particles of the steam press as forcibly upon each other as did the particles of boiling water. We find that heat is the repulsive force which keeps asunder the atoms of matter, as attraction is the force that draws them together. Between the two, the ever-changing, yet ever-constant, equilibrium of nature is preserved. The force with which the constituent atoms of a mass of matter press upon each other, measures the amount of heat felt. Where there is no matter, there can be no atomic pressure — hence no heat.

The pressure of Earth atoms upon each other gives a certain amount of sensible heat at the Earth's surface, independently of the vibrations caused by solar action; and because the pressure is increased as we descend below the surface, being equal to the weight of the particles at any point, added to the weight of the superincumbent matter, therefore the heat increases as we descend; the augmentation averaging about 1° for every 75 feet in depth. It is estimated that the Earth's surface is warmed 40 times as much by the Sun as by radiation of heat from her interior. In like manner the pressure of the Solar atoms upon each other is a measure of the amount of Sun heat. Let us compare the two:

We saw in the second lecture of this course

that the force of attraction at the Sun's surface is about 27 times that at the surface of the Earth; the Sun's particles being drawn together by the force of their mutual attraction 27 times more forcibly than those of the Earth. But we also saw that the density of the Sun is only one-quarter that of the Earth; the particles of the Sun are kept four times as far asunder, though pulled together with 27 times greater force than those of the Earth. The only possible explanation of this is that the repulsive force is 4 × 27, or 108, times greater on the Sun's surface than on the Earth; which is as much as to say that the Sun is 108 times hotter than the Earth. But 108 times 220 equals 23,760°, which should, therefore, be the heat at the Sun's surface, as measured from the absolute zero; or 23.500° above the freezing point of water. Like the Earth, the temperature of the Sun must increase, under greater pressure, with an approach toward his centre. The temperature of 23.500° is that at the point corresponding to the place where the air and water meet on the Earth.

At such temperatures as these it would seem impossible that matter could retain the solid condition, even for a moment; and indeed, we see that iron, and magnesium, sodium, and other metals, are reduced to a state of vapor in the Sun's chromosphere. But the spectroscope shows that they also exist in the liquid form nearer the centre, that condition being undoubtedly maintained by the greater pressure of the exterior matter. It is, perhaps, also possible that, near the centre, the still greater pressure may be great enough to form a solid nucleus for the Sun, notwithstanding the vastly augmented heat; as some of our gases can be compressed into the solid form. Of this we can know nothing; but we are enabled to judge that the Sun is, relatively speaking, a mass of fluid matter, only about one-fourth the weight of an equal bulk of earth, and lacking the solid crust which gives stability to the surface of our globe, and renders possible vegetable and animal existences, such as those by which we are surrounded. If there be any solid form of matter at any considerable distance from the Sun's centre, it is, in all probability, to be met with temporarily only, and near the outer limits of the solar atmosphere. Huge cakes, or flakes, of solid matter may be formed there, where the temperature is sufficiently low to permit it, just as ice is formed on the surface of a sheet of water; but the strong upward and downward currents, which we have yet to note at greater length, perpetually carry these flakes downward to be melted into liquidity by the greater heat; and their place is taken by cooler matter, to be, in its turn, solidified and carried down deep into the molten sea.

We have also reason to believe that the chemical elements existing in the Sun are not combined so intricately, or so numerously, as on our Earth. Here we very seldom obtain a pure element, except by artificial means, and then it is difficult to keep it pure. In the atmosphere we have a mixture of two principal gases — oxygen and nitrogen — but these are always found to be mixed with hydrogen and carbon; and, in addition to these, the atmosphere contains small portions of other forms of matter, either simple or compound. Pure water is a compound of oxygen and hydrogen, but we can seldom obtain it unmixed with other elements. So our rocks and earths are compounds; and even the metals have to be separated from baser dross, with which they are found associated, in the ore taken from the mine.

We know that on this Earth, up to a certain point, the elements combine most readily at high temperatures. But, far below the point of Sun heat this limit is passed, and the tendency of the different substances is to separate into distinct aggregations; and this appears to be the case in the Sun.

At the time of a total eclipse of the Sun, when the interposition of the Moon's dark body shuts off his light, large rosy-colored masses are seen projecting out beyond the edge of the Moon, which not infrequently attain an altitude of 50,000 miles or more above the Sun's normal surface. It has very recently been found possible to study these interesting objects in the absence of an eclipse. When the spectroscope is directed to the exterior portion of a protuberance, the only lines seen in the spectrum are those which indicate the presence of hydrogen; this shows that immense volumes of that gas exist as a distinct form, unmixed with other matter. Directed lower down, the instrument usually shows similar, but smaller, aggregations of sodium, magnesium, and iron. And these substances are not blended, as two gases will mix on our Earth, but exist separately, as oil and water will do when poured into the same vessel. The lighter gases appear to take the exterior place, just as they would do with us were it not for their tendency to unite chemically or mechanically whenever brought into contact.

These facts tend to prove that the Sun's ability to supply light and heat are not kept up by combustion, which involves a chemical union of previously separated elements. When we burn substances, we set in motion a very simple chemical process, in which the oxygen of the atmosphere unites with the hydrogen, or the carbon, of the burning body, and forms water, or carbonic acid; and this, if collected, and weighed along with the ashes, will be found to give a total exactly equal to the weight of the substance before it was burned, added to the weight of the oxygen taken from the air. And these results of combustion are unburnable. If the Sun were a burning body it would be necessary to get rid of tremendous quantities of water every minute in order to prevent the water from quenching the fire; and if the process could have been kept up, in spite of this difficulty, for the past 5,000 years, the Sun would by this time have been reduced to a mass of water, or a mere cinder, with nothing left to burn. But we have no evidence of the presence of water in the Sun, nor even of a single atom of oxygen, without which combustion can not be carried on. The only chemical combinations which appear to be possible are those on the Sun's exterior, beyond the limits of the luminous atmosphere; and the vibrations there excited are very probably the causes of many of the chemical activities which are produced on this Earth by the non-luminous portions of the solar ray.

It has recently been argued that the Sun's light and heat are kept up by the fall of meteoric matter toward him. It is well known that heat and motion are convertible into each other — that the sudden arrestation of motion produces heat, as shown in the ignition of meteorites by coming in contact with the Earth's atmosphere at the rate of about 30 miles per second. The exact equation of the two has been calculated; and it is found that a rise in the temperature of a pound of water, to the extent of 1°, is equal to the lifting of that pound of water through a height of 772 feet. It has also been calculated that, in the case of bodies falling into the Sun, the rate of movement just before the collision can not be less than 277, nor more than about 392, miles per second. Mayer computes from this that the fall of an asteroid into the Sun produces from 4600 to 9200 times as much heat as would be generated by the combustion of an equal mass of coal, and estimates that the aggregate of matter falling into the Sun every minute is equal to not less than 21,725, nor more than 43,550, cubic miles of water. According to this theory Mr. Thompson has calculated that the falling of the Earth into the Sun would generate heat enough to cover the expenditure of 95 years, and all the primary planets of the system would maintain it for only about 45,600 years. I have recently calculated that the quantity of matter coursing around the Sun, in the shape of comets, planetoids, meteorites, etc., is from one and a quarter to one and a half times as much as the aggregated masses of all the planetary bodies, including the Earth; and this would give a supply for about 65,000 years additional. Even with this addition, the Sun, as such, could not exist more than 110,000 years from the present time. But we have good reason to believe that the Earth has been warmed and lighted by the sun for a longer period than this, without any planetary accretion whatever. The theory is, therefore, untenable, and scientific men are generally abandoning it.

The only theory which seems to be compatible with the observed facts is that which assumes the solar phenomena of light and heat to be produced by the rapid and intense vibration of the Sun's matter, due to the great heat, which is, itself, principally caused by the tremendous pressure of the constituent atoms, resulting from the attraction of gravitation. The loss of this heat by radiation is necessarily followed, and partially compensated, by a continuous condensation of the solar mass, which would be appreciable at this distance, only after the lapse of thousands of millions of years. The telescope shows us that the Sun is the theatre of incessant movements, to the rapidity of which our Earth furnishes no parallel. The spots, which have been unusually numerous during 1870, are evidences of storms in the Sun's atmosphere, which are undoubtedly accompanied by tremendous convulsions in the interior mass. These spots, many of which are more than a million square miles in area, are shown to be depressions in the Sun's surface, almost funnel shaped, and are evidences of a downward rush of molten matter, similar to that which marks the whirlpools produced by two opposing currents of water; but on so much grander a scale that the comparison fails to convey even a faint idea of the phenomenon. The protuberances seen during a total eclipse of the Sun, and which appear at other times as brilliant protrusions, called faculæ, almost always visible in the neighborhood of the spots, are the swelling upward of the throbbing mass beneath, which would be paralleled on our Earth if a hundred volcanoes should vomit forth their liquid lava, all at the one time, and the accompanying convulsions of the interior fluid mass were unrestrained by the solid crust surrounding it. These spots and faculæ sometimes change their positions on the Sun's surface at the rate of 100 miles per second, and their forms change with equal celerity; spots having an area of many thousands of square miles, have been observed to form, and entirely disappear, within the space of a few hours. These mighty movements in the solar mass are cause enough for the vibrations needed to light and warm a million worlds like ours.

These are some of the lessons which are gathered by patient study of a single sunbeam. The ancient world, though not able to comprehend the value of light as a teacher of science, was yet wise enough to recognize it as a most potent instructor; and one of the most deep-lying fundamentals of the structure of human language is that which assimilates light to knowledge. Groping along blindly in the darkness of ignorance of nature's laws, the philosophers of olden times sighed for a flood of light to illume their pathway. We have attained that which "All their wise men waited for; and sought, but never found."

Strangely enough; modern science has illuminated the pathway of man into these arcana in a way directly opposed to ancient conceptions. Our predecessors prayed for more light; we found toomuch of it, for the purposes of study; and the grandest step in the investigation was taken when we had puzzled out the way in which we could slice it up into almost infinitessimal portions, and study in detail the dissected components of a single beam.

And it is thus with all knowledge. The human mind is so constituted that it can lay hold of but little at a time, though it can grasp immensity itself when taken in by little and little, while the mind grows with what it feeds on. Facts must be mastered in minute detail if the truth be attainable by finite minds. The knowledge of the infinite, like the light of the Sun, is too dazzling for the eyes of mortals; but the right acceptation of a simple revealed truth may help us onward in our path toward immortality; as we can reach out to a comprehension of the glory of the meridian Sun by grappling with a minute portion of his beams, and find in that knowledge a key to the conditions of other, and far more distant, worlds.

BIOLOGY OF THE UNIVERSE.

Habitability of other Worlds. Life on the Earth: Conditions of Organized Existence; Limitations of Temperature. Sun and Stars, too hot; Moon too cold—died out. Earth once like the Sun is to-day. Life on the Planets; Maps of Mars; the Planetoids. Invisible Worlds, Light, Heat, and Electricity; all Different Expressions of one Force. The Sun's Light and Heat variable; his Spots; their Influences on other Bodies; Earth weather of 1870. Variable Stars; Changes in Color and Brightness; Causes and Effects. Greater Cycles of Change. History of the Universe; Growth, Maturity, and Decay of Worlds, Systems, and the Universe. Diversity of Atoms. Future of the Cosmos.

One of the most interesting of all the questions which are suggested by a contemplation of Astronomical facts is: Are the Sun, Stars, and Planets inhabited? and, if so, are they, or any of them, inhabited by beings like ourselves, capable of reasoning out the leading facts in the conditions of the Universe?

The question has often been asked, and often answered; but the response was not always an intelligent one, though generally in the affirmative. We examine the Earth, and find that it is full of vitality—swarming with life, under all the conditions which seem to us possible. The surface is literally covered with animal and vegetable existences; and every one of those individuals, is, itsèlf, the seat of divided life; parasites innumerable live on its exterior, and countless millions of minute creatures exist in the interior, even of the healthiest human being. The waters of the ocean and river teem with life; and even the solid rock is in large part made up of the remains of animal and plant life, which enable us to trace out the history of our planet millions of ages ago. It would seem to have been the design of the Creator to multiply the individualities of existence on this earth, to the greatest possible extent — to cause the phenomena of life wherever they could be sustained, and to make of our world one vast theatre of vitality, in which every unit of living organism should not only enjoy life itself, but be the means of sustaining the lives of other beings. It has been argued that it is simply absurd to suppose that so much creative ability should have been exerted on our planet, while other and larger worlds were left unpeopled.

The argument from analogy is a very powerful one, when rightly applied; in fact it is the grand tool with which we work out truths in Nature — new to us, though older than the hills, which have been called eternal. But the argument is valueless, if loosely conducted. If we first study

*

out the conditions under which life is met with on our Earth, and find that those conditions exist in other bodies, we are warranted in assuming that organized beings exist in other worlds than ours. If, on the other hand, we find conditions in some other worlds which do not admit of the possibility of life, *as we understand it*, then we are warranted in stating the fact; though we can not say that there are no variations of material animation possible outside the limits of our sphere of observation. Of course we can only deal with material life. The domain of spirit existence is beyond our ken, and has nothing to do with the inquiry; since, in the case of spirit life, as distinct from material organization, there is no bond that connects it with this or that aggregation of matter which we call a planet or a star. We can not believe in spirits which are more intimately identified with Jupiter than with the Earth, unless as they may have been connected in the past with material organizations on, or in, one of those bodies.

We find that life is possible on our globe, under a wide range of conditions—in earth, air, water; as an independent organization, or growing and subsisting directly on some other animal or vegetable structure; or in animalcular form, inhabiting the globule of water, or the single drop of blood, in uncounted millions. There is only one mode in which material life can be gauged universally; that is by the fact of organization; there is only one way in which the possibilities of organization can be limited: — by temperature. And we find that almost every class of animal and vegetable life has its own very narrow limits of temperature, ranging over but a very few degrees of the scale, while all are possible only within a few score degrees of temperature. Hence it is that most animals and plants are met with only within narrow belts of longitude; the reindeer and seal of the polar regions could not change places with the tiger and alligator of tropical climes; the grape requires a warmer climate than the apple; barley will grow where it is too cold for wheat culture; and wheat is grown successfully in climates where the solar rays are not strong enough to mature a single ear of corn. It seems as if every separate organization recognizes one particular degree of heat as that which is most congenial to the development of its structure; just as it appears probable that the properties of the several chemical elements are also determined by the different capacities for specific heat, which have been impressed upon their atoms.

The domain of animal and vegetable life, as ordinarily understood, has a range of only about 90° of temperature, or from 30° to 120° of the Fahrenheit scale. At 32° water (a large component of organized fluids) congeals, rendering impossible the circulation of blood in the animal, and of sap in the vegetable; while above 120° the chemical affinities are so active as to forbid permanence of form. The animal world is provided with apparatus for keeping its temperature nearly equable, at an average of about half way along this scale. Man and the warm-blooded animals keep in the upper half of the scale by dint of burning large quantities of carbon within the body, the carbon being taken up as food, and the required oxygen being taken in through the lungs. And the amount of heat thus evolved measures the muscular power of the animal, as of the steam engine. The total muscular force exerted through the life-time of an ordinary man is about equal to the force which is obtained by burning three tons of coal.

The same process of heat-making goes on, but much more slowly, in the colder-bodied fishes and reptiles. Hence the necessity for an atmosphere — either of air, or water, or both; and this is also necessary to give force to the solar ray, as explained in the preceding lecture. Without an atmosphere, the Sun's rays would not warm the surface of our globe sufficiently to keep up the temperature required to support life; even if it were not necessary to respiration.

All life is not impossible outside of these narrow limits of temperature; but the range is yet small, within which living organisms can exist.

We may lay it down as a rule that the higher the form of vitality the more complex is the accompanying organism, and the less the ability to depart from the normal standard of temperature. It is true that man can endure great extremes of heat and cold; but this is accomplished only because, artificially or otherwise, he is able to preserve the healthful temperature under adverse conditions. You can not change the temperature of the human *body* 2° either way without causing violent sickness; nor 10° without producing death. But some of the lower orders of organized existence have been imbedded in ice, and others immersed in water nearly at the boiling point, without losing their vitality. The whole range under which any form of organized existence is possible, may be safely estimated to lie within 300° of the Fahrenheit scale; between 100° below and 200° above, the freezing point.

The temperatures at the surfaces of the Sun and fixed stars are far above these limits — can be measured only in thousands of degrees of heat, instead of hundreds. And the heat is so great as to prevent the matter of which they are composed from retaining the solid form. Hence there is no more of fixity on the surface of a self-luminous star than in mid-ocean; and the idea of special locality is absent. We conclude, therefore, that the Sun and fixed stars are not habitable globes themselves; but that their office is to furnish the light and heat necessary to make other globes habitable. They are like the Shaksperian hero; who, though not witty himself, was the cause of wit in others. The analogy does justify us in concluding that the numberless fixed stars which dot the firmament were not created in vain, and that they are, therefore, like our own Sun, the centres of vitalizing power, each to many worlds, though so far removed from us that we see them not; the light of the stars, reflected from their surfaces, is too feeble to reach our eyes at such immense distances.

How is it with our nearest neighbor, the Moon? In the first place, she has no atmosphere. We never see any clouds in the Moon, even with the most powerful telescope; the same dull, unvarying appearance is always presented to us, except as her change of position with regard to the Sun enables us to see more or less of that side which is always turned towards us. The Moon has very often been seen to pass, apparently, over a star, hiding it from view for about an hour. If the Moon had an atmosphere, the star would appear and disappear gradually, being comparatively dull when seen through the aerial envelope of the Moon, just before, and just after, the apparent contact with the solid bulk of our satellite. But the appearance and disappearance are instantaneous. This fact also proves that there is no water on the Moon's surface; because water will always vaporize, even at the lowest temperatures, if the pressure of the atmosphere be removed. Place a basin of water, or even a lump of ice, under the receiver of an air pump, and exhaust the air. The glass will be immediately filled by a cloud of vapor rising from the water or the ice. If there were any water on the Moon it would form an atmosphere; of which there is no trace. The Moon has no aerial envelope, and therefore no breathing animal can exist on her surface; nor can the ray of sunlight produce warmth enough to support the existence of any vital organism. There is no life, no cloud, no wind, no water, no refraction of light, on the Moon. Her surface is a barren desert, from which both Sun and stars could be seen at the same time against the black background of the sky, could any being view the scene from such a situation.

But it was not always thus with the Moon. She was not ever the barren waste we see her now. Her surface is diversified by bright prominences, which are known to be mountain peaks and ranges, and hill country; broken up by volcanoes, all of which are believed to be now extinct. These mountains are much higher, proportionately, than any on our globe; some of them being 24,000 to 27,000 feet high. Not less than 39 peaks have been measured, the height of each of which exceeds that of Mont Blanc. Between these mountains are dark patches, that were formerly thought to be oceans, but the large telescopes recently come into use show them to be nothing but old sea bottoms. They were once filled with water, and the Lunar surface was once diversified by land and ocean, lofty mountain scenery, and fruitful valleys, as the Earth is now; and was surrounded by an atmosphere which permitted the existence of beings similar to those found on this Earth. But the water has vanished, and the atmosphere has departed, making the Moon a vast burying-ground, on which repose, in undisturbed stillness, the *debris* of untold generations of animal and vegetable life.

In the Sun, Earth, and Moon, we have three remarkably distinct sets of conditions, which it will be instructive to compare. The Moon has once been habitable like our Earth; has the Earth ever been an incandescent mass, like the Sun? and will both Sun and Earth ever be reduced to the present condition of the Moon?

To the first question we can answer "Yes;" to the second — "very probably."

The Earth becomes warmer, at the rate of about 1° for every 75 feet, as we descend below the surface; and fully 300 volcanoes, on different parts of the Earth's surface, vomit up red-hot liquid matter from her interior in the present century — two of the most active of which are situated in the polar regions. We refer to

Mount Hecla in Iceland; and Mount Erebus, near the South Pole. We have every reason to believe that the Earth's solid portion is but about 16 miles in thickness, and that all within this is a densely fluid mass, 6 or 7 times heavier than an equal bulk of water, in an intensely heated condition, analogous to the state of the Sun. The present crust of the Earth has been made by the process of cooling, through a period of not less than 100,000,000 of years, previous to which it was undoubtedly a mass of incandescent matter, and emitted light and heat as the Sun does now. That the whole globe was once fluid is proven by the fact that the Earth's polar diameter is 26½ miles shorter than the Equatorial diameter — a fact which can be accounted for in no other way than on the supposition that she then rotated on her axis, as now; and that the attraction of the matter in the equatorial regions, being partially overcome by the velocity of rotation, the yielding mass accumulated toward the Equator, causing the bulging shape which is shown by the measurements spoken of in the first lecture of this series. Take a ball of dough, and thrust a stick through the middle, then turn it around rapidly; it will be found to assume the flattened shape noticed in the case of the Earth. If, however, the dough be stiffened, by baking, the change in shape is no longer possible. The Earth could not have become flattened at the poles unless once fluid; nor after the formation of a thick crust by cooling.

The former incandescent fluidity of the whole Earth is more directly shown by an examination of its cooled crust. We find it to consist of a number of distinct layers, which, although somewhat mixed in many places by subsequent upheavals of the molten mass beneath, yet occur with sufficient regularity to permit us to ascertain the exact order of their formation, though we may not be acquainted with the first in the series. We find that the lowest of the known rocks, the granitic, are of fiery origin; though some of them have been thought to show signs of stratification. They are composed of nearly one-half oxygen, and the other half a fused mineral. The rocks above the granitic have evidently been formed from it, and each other, by the action of water; and have been deposited, one above the other, in layers, which contain the remains of animal life, showing that some forms of organized vitality have existed on the Earth ever since the first formation of a granitic crust, of perhaps not more than half a mile in thickness. And not only the individuals, but the types of these existences, have been many times changed in the past, to correspond with the altering conditions of the surface.

The Earth is still cooling; but the process is less and less rapid as the increasing thickness of the crust augments the resistance which it offers to the escape of the internal heat. Bischoff estimates the present rate of cooling to be about $1\frac{1}{2}°$ in a million years, and that the internal heat adds only $\frac{1}{30}$ of a degree to the temperature of the surface. As it cools, the interior mass is continually shrinking in bulk, leaving an interval between it and the crust. If the outer shell were of uniform weight and consistency, its stability would be but little affected by the shrinking. But it consists of substances unequally cohesive, and gives way in the weakest part, when the whole crust closes in around the central mass. These collapses would be frequent in the early days of our geologic history; they occur now at longer intervals, as the crust becomes stronger with increasing thickness. We have evidences that several such collapses have occurred in the past. But a time at length arrives when the crust is so thick that it is strong enough to bear the strain of its own weight. The crust will also continue to cool as the internal heat diminishes, and will crack open with the resulting contraction in its bulk, forming huge fissures above, and vast caverns beneath, into which the waters of the ocean will rush, and thence be absorbed by the chilled strata, losing themselves for ever out of the reach of the evaporating forces now at work on the surface. It is estimated that our Earth's crust has already absorbed one-fiftieth part of her original ocean, and that every drop will have disappeared by the time the crust has attained a thickness of one hundred miles. The Earth could absorb 50 times as much water as now exists on her surface, if cooled down to a sufficiently low point. When the water disappears, the phenomena of evaporation and rainfall, which now keep up the circle of life sustenance here, will vanish also. The cold atmosphere will lose its vapor of water, as the capacity of the air for taking up vapor diminishes rapidly with decrease of temperature; and the atmosphere itself will at length disappear, its component gases becoming too cold to preserve their elasticity. Long before this every living thing will have died out from our globe, and its surface will be left as barren and as inert as that of the Moon is to-day.

It is now very well understood that the capacity of the Earth to sustain organized creatures is being rapidly diminished in another way. Carbon, as one of the four organic elements, is essential to their existence, and they are continually engaged in eliminating it from the air and from water, one important product being the carbonates of lime (as chalk), which form a large proportion of the bony structure of animals. This compound is resolved with difficulty, and it is now a well-established fact that the deposits of carbonate of lime are rapidly increasing, especially at the bottom of the sea, where they lie undisturbed by the action of the waves above. The atmosphere is thus losing its carbon, and therefore losing its ability to supply the material from which are built up the bony tissues of the animal economy. Long before the atmosphere is absorbed by the Earth's crust it will have lost all its carbon, and lose with it the capacity to supply the skeleton which is the physical foundation of animal life.

We have telescopic evidence that the Moon has passed through the changes above noted. As many as 425 lines have been counted on her surface, which are called *rilles*, and are undoubtedly cracks, of the character referred to, through which the waters and the atmosphere have disappeared. The Moon has passed through these destructive phases much sooner than the Earth, because she is so much smaller a body, and presents a much larger radiating surface in proportion to the volume of matter to be cooled. The Earth's diameter is (3.666) $3\frac{2}{3}$ times larger than that of the Moon; her surface is $13\frac{1}{2}$ times greater, and her volume is 49 times greater. Hence, the Moon's radiating surface is, proportionately, nearly four times greater than that of the Earth.

In this difference of mass we have, also, the cause of the slow rotation of the Moon on her axis — only one rotation being performed in a little less than a month. The attraction of the

Moon causes a heaping up of the waters of our ocean, to the height of about four feet in the open sea, forming the tides, which follow the Moon in her apparent diurnal passage around the Earth from East to West. This being in a direction contrary to that in which the Earth is turning, the result of the friction of the waters is to retard the Earth's time of diurnal rotation, at the rate of about 1 second in 4½ centuries. When the Moon had an ocean the Earth caused corresponding tides on her surface, but nearly 85 times greater, the Earth being *so* much heavier than the Moon, and the retarding force was correspondingly increased. The force of rotation on the axis, relative to the Earth, was destroyed by the friction, before the waters left her surface; and the effect of the Moon's attraction on our ocean will be to destroy the Earth's axial rotation in precisely the same way. The time required to work this stupendous change is estimated to be about 26 millions of years. This retardation of the Earth's axial motion, combined with an actual acceleration of the Moon's movement in her orbit, gives rise to an apparent quickening of the Moon's rate of movement round us, as measured in Earth days, to the extent of about 12 seconds in a century. It is necessary to make this allowance in order that the calculation of former Eclipses shall agree with the recorded observations of the times when they occurred.

For the same reasons as those we have given, the Sun has not cooled off so rapidly as the Earth, when at the same actual temperature, though he is losing heat more rapidly than the Earth is now. His volume is 1,250,000 times that of the Earth, while his surface is only 11,634 times greater. Taking the estimate of Professor Mayer, that the Sun's surface cools now at the rate of 3¼° annually, he would still give out light and heat enough to maintain life on our globe for 60,000 years to come; and it is probable that this estimate of time is much too small. I have calculated that the Sun would lose one-quarter of his present lighting and heating force in about one hundred thousand years, by radiation; if there were no compensation by compression, or the fall of planetoidal matter to his surface. We can scarcely suppose it possible that the Sun should be lighted and heated by some far-distant body, sufficiently to make life possible on his surface after his own light has gone out forever. The Sun was never inhabited; and probably never will be, otherwise than by beings who can exist without light. He is still fluid, but, unlike our globe, he presents no evidence of flattening at the poles. The reason is that his velocity of rotation is relatively small. He occupies 25⅓ days in turning once round on his axis. The rotation of the Sun lessens the weight of bodies at his equator by only one part in about 18,000.

At the Earth's equator the loss of weight is more than one part in 300.

When we examine the light of the planet Jupiter we find that it is *reflected* light, but so intense in proportion to the area of reflecting surface, and the distance from us, as to lead us to conclude that he reflects *all* the sunlight received by him, which he could not do unless his atmosphere were very dense — almost impervious to the passage of the light ray. Looking at him through the telescope, we see his apparent disc crossed by broad bands of dusky hue, which cross a brighter background, in directions nearly parallel to the equator of the planet. These belts vary their form and position, showing that they do not form a part of the solid planet, like the unvarying irregularities of the lunar surface, but that they are phenomena of his atmosphere, like the spots in the Sun. If we compare them with the currents in the atmosphere of our Earth, we find a striking similarity of position and change. On our globe we have the trade winds blowing from the East, on each side of a calm-belt, which shifts back and forth across the equator, with the changing declination of the Sun. Outside of these we have a calm-belt in each hemisphere, and further towards the poles another belt of winds, called the anti-trades, which blow from the west. We see also spots among his belts, analogous to the Sun spots. We conclude that we do not see the planet at all, but only his atmosphere, which is agitated in a similar manner to that of the Earth, but is of much greater density, hanging around the planet like a heavy cloud, and containing in suspension all the water which will form the future oceans of Jupiter. The planet is some 1300 times larger than the Earth; but, if the disc we see be that of the solid body, its density is but (0.22) a little more than one-fifth that of the Earth — scarcely so great as that of the Sun. Either Jupiter is as yet uncondensed into the solid form, or his crust is not sufficiently cooled to permit the deposition of water on his surface from the vapor clouds in his atmosphere. In either case he is not yet fit for the habitation of living beings; and this accords with the analogy that, being so much larger than the Earth, he should be much longer in cooling, while he should have cooled more rapidly than the Sun, as being less in size than the central luminary. Jupiter was a self-luminous body, a sun to his four moons, not many thousands of years ago; he will be a habitable globe, like our Earth, many thousands of years hence. His satellites are probably habitable to-day. All but the second are larger than our Moon, and the diameter of the third is nearly one-half that of the Earth. The spots seen on the surface of Jupiter show that the vast orb of 85,390 miles in diameter rotates on his own axis once in a little less than 10 hours. This great velocity causes a corresponding flattening at the poles, of $\frac{1}{11}$, while that of the Earth is only a little more than $\frac{1}{300}$.

Saturn is ninety times as large as the Earth, and his density is only about $\frac{1}{8}$ (0.12); furnishing an argument similar to that found in the case of Jupiter, which is strengthened by a parallel observation of atmospheric belts. We consider it probable that Saturn is in a more advanced stage of planetary development than Jupiter, but that he is as yet uninhabitable; and that his moons are the only members of his system which are at present capable of sustaining the higher forms of life upon their surfaces. The condition of Saturn is probably like that of the Earth in the era when our carboniferous rocks were being deposited from the atmosphere and the water, which were thus purified for the use of the higher animals in the next geologic era. Under such circumstances it is useless to speculate on the views obtainable from the surface of Saturn. The inhabitants of that planet in the far distant future, when the Earth has chilled down into majestic death, will doubtless see the satellites, and perhaps others that have escaped our notice; but they may know nothing of the magnificent rings which puzzled our astronomers for so many years. Those rings will, in all probability, have broken up to form one or more additional moons, before the advent of highly organized beings on the surface of Saturn. Struve is said to have recently seen the obscure inner portion of the ring settle down upon, and spread over, the body of the planet.

We find on the planet Mars indications of conditions very similar to those obtaining on the Earth's surface; though differing widely in many important particulars. Mars is nearly of the same density as the Earth; though, being less than 5000 miles in diameter, his volume and weight are but about $\frac{1}{8}$ part of that of our globe. We may infer from this that he has cooled a little more rapidly than the Earth, especially as he receives less heat than we do, being farther from the Sun. But we find indubitable evidences of land and water, and snow on his surface, and clouds in his atmosphere. His poles of rotation are seen to be capped with ice or snow, which increases or diminishes in bulk as the pole is turned to or from the Sun. Hence the processes of evaporation and rain-fall, melting and thawing, are perpetually going on; as in the case of the Earth. The melting at the south pole of the planet, in particular, has been observed; and it occurs rapidly, that pole being turned toward the Sun when the planet is nearest to him, and also nearest to the Earth's orbit. Mars was in this position in 1862; and was so near to us as to permit the principal features of his surface to be mapped out by Mr. Dawes. It has seas and continents like ours, and yet unlike in their general distribution. The accompanying maps of his two hemispheres, which I prepared a year ago for the "Chicago Tribune," show the distribution of land and water on the surface of Mars. The most prominent of these have been named, and the letters refer to the names our astronomers have given to the more important continents and seas. We have no means of knowing how they are designated by the students of Erotography (*Eros*, Mars) who live on the planet.

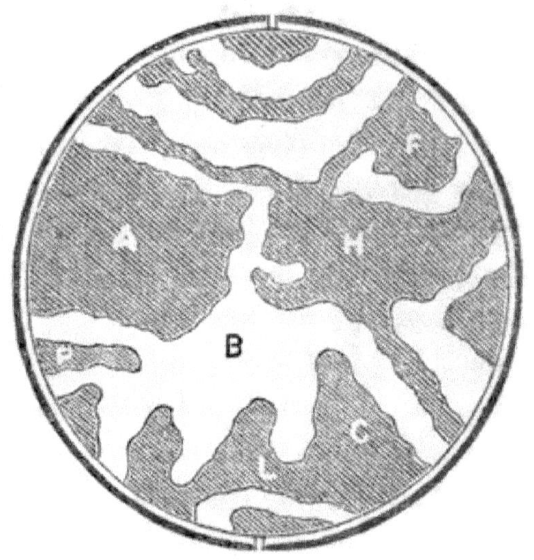

The shaded portions of the diagram represent the land; the white spots show the relative distribution of the water. The upper part of each diagram represents the north, and the breaks in the bounding circles show the positions of the poles. Each pole is surrounded by a patch of ice, marking the middle of the polar regions. The following are the references:

A. Dawes Continent. C. Cassini Land.
B. Dawes Ocean. D. Delambre Sea.

E. De La Rue Sea. L. Lockyer Land.
F. Fontana Land. M. Mädler Continent.
H. Herschel Continent. N. Tycho Sea.
J. Maraldi Sea. P. Phillips Island.
K. Kepler Land. R. Secchi Continent.

A comparison of the above diagrams with a hemisphere map of our world will show several important differences. On the Earth the land lies in compact, though irregular, masses, and the tides of the ocean have free course, except where they pass between islands, or through channels that separate islands from the main land. The surface of Mars is marked by numerous seas of the bottle-neck form, and these run between continents and lengthy peninsulas; only one rounded island being visible, and that is probably a volcano. On Mars the land and water are nearly equal; while on the Earth the waters cover nearly three-fourths of the entire surface. The land of Mars is of a ruddy color; and it is the reflection of

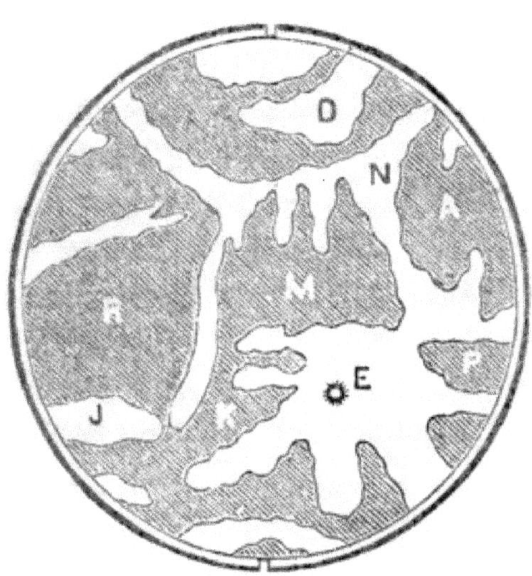

the solar rays from this which gives the red appearance of the planet. The water is of a greenish hue (as seen through the telescope), and the latter fact indicates a condition similar to the waters of the Earth, which are blue or green according to distance from the shore. The apparent hue of the waters is undoubtedly modified somewhat by the passage of the rays of light through the Martial atmosphere, before they pass through "the ethereal void" to impinge on the aerial envelope which surrounds our Earth. Mars presents a very large extent of coast-line as compared with the Earth, and it is apparent from the diagrams that it is possible to travel by land to almost every part of his land surface without resort to navigation.

Mars is, therefore, adapted as a residence for rational beings, like ourselves; and it may be that they have attained to a higher stage of mental development than we have, for the double reason that the planet appears to have been habitable some thousands of years longer than our Earth, and also that the extremes of temperature are greater — the latter stimulating inventive ingenuity, as in a lesser degree with ourselves. The length of the Martial day is about the same as ours, and the inclination of his axis to the plane of the orbit is nearly the same as in the case of the Earth; but the eccentricity of his orbit is so great that if his nearest distance be represented by 5, his greatest distance from the Sun will be 6, and he receives one third more heat and light in the former position than in the latter; while the Earth differences of temperature are but as 14 to 15. The annual range of temperature at any one place on the planet Mars will be one-fourth greater than with us; and the equalizing effects of ocean currents must be very much less felt on that planet, owing to the peculiar distribution of land and water on his surface, and the comparative absence of ocean tides — Mars being unattended by a moon. Hence Mars has much less of atmospheric disturbance than we have; the seasons shading off gradually, from hot to cold, and from cold to hot again. But the amount of solar heat, and solar evaporation, must have a much wider range than on the Earth, owing to the great eccentricity of the orbit of the planet. This will also give rise to greater differences of extreme and average temperature in the two hemispheres than with us; and, hence, give a possibility of far wider range, and much greater differences in the modes of existence, than are known on the Earth. From this we may infer that the number of classes and species of vegetables and animals is much greater, while the number of individuals of each order is smaller, on the surface of Mars than on that of the Earth; and it is very probable that the majority are monennial — having an existence limited to one year of Mars life.

Locomotion on Mars would be very easy to men possessed of our muscular strength, as the force of gravity is only about two-fifths of that at the Earth's surface; and the loss of power by friction must be correspondingly reduced. If the inhabitants of Mars have no moon of their own, they can see ours with the naked eye, if blessed with vision as keen as ourselves; they have probably never seen the planet Mercury, and may be able to see the larger Sun spots, but not the solar corona, being strangers to the phenomena of an eclipse. They may catch occasional glimpses of Jupiter's moons, and see several of the planetoids, where we have never caught a sight of any other than Ceres and Vesta without the aid of the telescope.

We have already indicated that Mars has nothing to compare with our tidal phenomena, owing to the absence of a moon. With us the attraction of the Moon is to that of the Sun as 51 to 20. Hence, the magnitude of our tides being represented by 7, that due to the Sun would be 2. The greater distance of Mars from the Sun, and the lesser distance of his waters from the centre of the planet, will reduce the height of the tidal wave in his larger oceans to three or four inches; while the tide in our open ocean is four feet. The rush of waters through the narrow inlets will be somewhat greater, though slight — just enough to keep up a moderate circulation in the waters. With such a land contour on our Earth the greater tides would cause rapid changes in the plan of the continents. In Mars we have no very active cause of mutability in this direction. The permanence of such an outline as is presented by Mars would be practically impossible were he attended by a moon of considerable weight. Here we have one among numerous examples of "the eternal fitness of things" to their surrounding conditions.

In this lesser tidal flow, we have also an absence of the forces which have produced such great changes in the Earth's surface, in cutting channels through what was once an isthmus, and the separating an island from the mainland. But, inasmuch as Mars exhibits a greater propensity for channel-forming than is indicated in the case of our Earth, we can but conclude that the original volcanic action, which elevated the land masses above his mean surface, operated largely in *lines* of force, whereas the upheavals of the Earth were often effected in *points*; as is attested by our numerous islandic formations. And, strangely enough, these elevating forces appear to have operated in the polar regions nearly parallel to the plane of the equator; while in the equatorial regions these lines of upheaval are more nearly perpendicular to the direction of rotation on the axis. This irregularity of formation is undoubtedly due to the great eccentricity of the orbit of Mars; the more important ruptures in the once thin crust, occurring near the time of the perihelion.

It is also noteworthy that the positions of the ice formations indicate that the poles of Mars are the regions of greatest cold; which is not the case on the Earth's surface, the point of minimum temperature in our Northern hemisphere lying ten or twelve degrees from the pole, toward the American Continent, while our magnetic poles show an equally wide departure from the poles of rotation. It is already known that the position of the magnetic pole is deducible from a study of the lunar motion; and it is highly probable that the positions of the points of least temperature will yet be traced to the same cause. Mars being unattended by a moon, there is, in his case, no apparent cause for difference in the average locality of the several poles of rotation, magnetism, and temperature.

We know much less of Venus than of Mars; but astronomers have discovered evidences of a dense cloudy atmosphere; and irregularities on the outline, which are believed to indicate the existence of mountains twenty miles high. Venus differs but little from the Earth in size or density; and her day is about the same length as ours, giving a very small compression at the poles. Venus is undoubtedly habitable. She receives about twice as much heat and light from the Sun as we do, but her surface is not, necessarily, much hotter than that of the Earth, as her atmosphere may be adapted to the increased outer heat; and, even if not so, the temperature is still below the upper limits of material organization. Her atmosphere has been computed to be one-fourth more dense than ours; and this agrees with the probability that the greater nearness to the Sun produces much more evaporation from the ocean than with us, and causes the peculiar clouded appearance of her atmosphere, which at once shuts out the direct rays of the Sun, causing him to be seldom visible, and prevents us from obtaining a knowledge of the distribution of land and water on her surface. Venus has no moon, and therefore no tides, in our sense of the term; but her axis being inclined fully 55° from the axis of her orbit, her inhabitants are subjected to great irregularities of temperature in the different seasons, with great differences in the lengths of days and nights. And the changes of position ensuing from this phenomenon, bringing widely different parts of her surface under the Sun in rapid succession, must cause rapid variations of temperature as affected by atmospheric currents. The evaporation being so much greater, the trade winds must blow with hurricane velocities, and the general positions of the belts vary rapidly and widely;

while rain-falls must be frequent, spasmodic and heavy, causing swift-flowing rivers, and stupendous changes of land outline, accompanied by countless earthquakes and numerous volcanic eruptions. We may guess the inhabitants of Venus to be short and stout, as befits those who have to do constant, sturdy battle with the elements; and to be possessed of little knowledge of the great universe without, owing to the difficulty of seeing through an atmosphere always surcharged with vapor. Hence they may know but little of the exact sciences, which, among us, have been cultivated largely in deference to the needs of astronomical research. We may infer also that their fruits and cereals are large and juicy; and that the times of seeding and harvest come within an earth month of each other; both being performed during that short interval in their brief summer, between the deferred spring gales and the early autumn tempests.

Mercury is probably too near the Sun to be habitable by any but a very low order of beings; especially as the great eccentricity of his orbit causes a change every six weeks from a position where he receives ten times as much heat and light as the Earth, to a point in the orbit where he receives only 4½ times as much heat and light as ourselves, or the reverse. It has been graphically stated that this change of temperature is nearly equal to the difference between melted lead and frozen quicksilver; so that, whatever may be the absorptive properties of his atmosphere, the change is probably too great to permit life of any but a low order on his surface. His atmosphere is very dense; and it is not improbable that Mercury is not yet cooled sufficiently to permit the formation of a solid crust, though some observers claim to have seen a mountain eleven miles high in his southern hemisphere.

Of Uranus and Neptune we know so little, owing to their great distance, that we are unable to form an estimate of their present adaptability to organic life on their surfaces; though the analogy justifies us in believing that they *have* been, are, or will be, the theatres of organized life. We can compute that, with the same atmospheric medium as that which envelops our Earth, Uranus only receives one part in 368, and Neptune only one part in 904, as much heat and light as we receive from the Sun. But the atmospheric conditions in each case may be so ordered, that each receives enough of light and heat to sustain life on his surface; for we must remember, as explained in the preceding lecture, that the intensity of the sensation is dependent upon the medium through which the force acts. If so, the view of the heavens from Uranus will be moderately interesting, as his sky is lighted up by four moons, though they can not shine so brightly as ours; but the out-look from Neptune must be blank indeed. The inhabitants of the outermost planet may be able to determine the parallaxes of a great many more of the fixed stars than we can; but to them the Sun will be a very uninteresting looking object, scarcely twice as large as Jupiter appeared to us last winter, and none of the planets in the system, inside the orbit of Jupiter, will be visible. To the inhabitants of Neptune, our Earth, with its history of overwhelming interest to us, its innumerable scenes of joy and sorrow, its grand achievements in the realms of science, will be too insignificent to be noticed from its oldest brother. Well might the Psalmist exclaim — "What is man, that *Thou* should'st be mindful of him?"

We may dismiss briefly the planetoids which revolve between the orbits of Mars and Jupiter; the number of which is anywhere from 112, known, to 150,000, yet to be discovered. The largest of these bodies — Ceres and Vesta — are scarcely more than 280 miles in diameter, and would cool down to the zero of life in $\frac{1}{90}$ part of the time required for the Moon to become uninhabitable. Of the smaller ones there is still less necessity to speak; they were all, probably, used-up planets, thousands of years ago; and life was permitted on the surface of one of those bodies but for a very brief period. It would be highly instructive could we trace the record of one of those little worlds through a thousand years of earth time; it would present an epitome of the history of our planet for many millions of years. But it is a sealed book to us; except as we can spell out the few characters on the exterior of the volume. No mortal eye may scan the hieroglyphics themselves — much less be permitted to decipher their hidden meaning, and add its lore to the fragmentary fund of human knowledge.

Still less are the aerolites inhabited; no life is sacrificed to the stern fiat of the law of attraction, in the case of the millions of aerolitic bodies which fall to the Earth yearly, unless we except the case of one of the ancients, whose name has escaped me; he is reported to have been struck dead by a stone falling from mid-air, which was probably an aerolite, though said to have been dropped by an eagle. He is the only man ever known to have come into collision with another world; and the shock was fatal. We have still less reason to think it possible that the cometary masses are habitable. Their gaseous constitution, and the terrific alternations of heat and cold to which they are subjected in a single revolution, forbids the supposition. The nebulæ are uninhabitable, like the comets; affording no foothold for organized existence.

We see, then, that of the untold millions of objects which the telescope shows us exist in the firmament, only a very few — a number small enough to be told on the fingers, at twice — are probably habitable by any beings which require the organization of material atoms, as a medium for the exhibition of vital powers and functions. How vast must be the number of invisible worlds which *do* sustain life, in order to carry out the analogy suggested by a look at our own globe, it is impossible to tell; or even to guess. But of one thing we may feel assured. What we do see, even with telescopic eye, is not more than a drop in the bucket, as compared with the boundless ocean of existence that we can not see. Every leaf around us is teeming with life. Every pulsation of the heart sends coursing through our veins millions of creatures, all organized to be, to do, and to suffer — genuine verbs of existence. Every mouthful of water we drink, or food we eat, involves the destruction of thousands of living beings; and with every breath we draw we change the current of life to many an organism. This world is a great theatre of vitality; and we can not doubt that the number of such theatres of action is as great as, or greater than, that of the centres of light and heat; which seem to have been established for the express purpose of enabling those unseen worlds to fulfill their mission.

We are thus enabled to infer the conditions of existence on other worlds to-day; but we have also seen that the conditions obtaining on any one body are dependent on those of other aggregations of matter. Changes in one produce effects on all. The attraction of the Earth has destroyed the axial motion of the Moon; and the rotation of the Earth is slowly dying out as a consequence of the Moon's attraction on the waters of our ocean. Looking out still further, we find correspondent influential changes in other bodies. The material particles which make up the mass of the Sun are in a state of incessant vibration; and these vibrations give rise to the phenomena indicated by the names, light, heat, and electricity, which all appear to be but different expressions of the same force, according to the medium through which it acts. We can gain an idea of this convertibility of function by exciting to rapid motion the atoms of a mass of matter; either by friction, as in the electrical machine, or by inciting chemical change, in the galvanic battery. If we allow the excitement, thus produced, to pass along a piece of wire, which offers no resistance to the flow, we have an electric current. If we interpose what is called an imperfect conductor, the flow of the current is partially arrested, and gives rise to the sensation of heat. If we cause the current to flow from one point to another, through space, without affording a conductor, it produces light. The force which is evolved by the Sun particles, in moving among each other, is electricity at his surface; traveling through the void, which presents nothing but an extremely tenuous ether as a conductor, it is light; communicated to us by imperfectly conducting matter in the atmosphere, it gives heat.

And we find that this force varies, not only in the form of its expression, with the medium through which it acts, but varies also in its intensity, in sympathy with the positions of other masses of matter. The changes which mark the seasons are caused by the varying position of the Earth's axis; and the minor changes, occurring almost daily and hourly, are due to altering conditions of our atmosphere. But the force of the solar light and heat is also variable. The spots, which we saw in the preceding lecture are evidences of violent storms in the Sun, have a regular cycle of about eleven years, corresponding nearly to the revolution of Jupiter. Thus: a very few spots were visible in 1856, and the number increased in 1859 and 1860; observations being made of 297 in the latter year. From this time the number decreased for about five years, and again increased to 1870, when the number seen was greater than at any time during the century. Jupiter is, by far, the largest planet in the system; and it is but natural to suppose that he produces the greatest effect. But Venus is also found to exercise a powerful influence in the formation of spots; they being always most numerous when Venus and Jupiter are on the same side of the Sun, both operating in the same direction. It is also considered probable that there is a greater cycle of 58 to 60 years, corresponding to five revolutions of Jupiter and two of Saturn; and a still larger cycle of 168 years, measured by two revolutions of Uranus and one of Neptune. As the maximum times of these different cycles coincide or interfere with each other, the relative extent of the Sun storms will vary; just as the height of the tides in our ocean is greatest when the Sun and Moon are acting together (on the same line), and least when their attractions are pulling at right angles to each other. I pointed out the fact that near the time of the maximum exhibition of Sun spots in the Spring of 1870, the Earth, Sun, Jupiter, Mars, and Venus, were all nearly on one straight line in the heavens.

These planetary movements cause storms in the Sun, just as Earth storms and tides are produced by the varying positions of the luminaries; and can be calculated on with certainty beforehand. The astronomical world had long known that 1870 would be a maximum year.

And these changes, though great, fully one part in 130 of the whole apparent surface of the Sun being covered by black spots last year, seem to be but small compared with those which have occurred in the past. Abul Farajius, an Arabian historian, tells us that in the 17th year of Heraclius, half the body of the Sun was eclipsed, continuing for eight months; this corresponds to the year 626 of the Christian Era. It is more than probable that the great darkness at the time of the crucifixion was caused by an enormous Sun spot; as there are grave difficulties in the way of supposing it to have been produced by an eclipse of the Sun.

We find similar changes, in the present age, in the fixed stars. The periods of more than 100 variable stars are known; and there are many others known to be variable, the periods of which have not been ascertained. The star Algol (Beta Persei) changes from the 2½ magnitude to the 4th, and back again, in 2.9 days. It shines as a star of nearly the 2nd magnitude for 2 days 13½ hours; then loses its light for 3½ hours; then increases in brilliancy for 3½ hours, and regains its former brightness. It is now believed that this phenomenon is due to the passage of an opaque attendant across his disc. But the star Mira (Omicron Ceti), or the Marvelous, has a period which can not be accounted for in this way; and so have very many others. Mira is usually seen as a star of the 2nd magnitude for about 15 days, then decreases in light for three months, at the end of which time it is only of the 11th magnitude, and can scarcely be seen for five months, even in a good telescope. It then increases in brilliancy for another three months; the whole cycle having been performed in 331⅝ days. This period is subject to variation, and so is the amount of light received from the star, by us, at different times of maximum; it is sometimes no brighter than a star of the 4th magnitude, when at the brightest phase of one of its periods. In this star, and in many others of the same class, we have very probably a similar set of conditions to those existing in the case of the Sun, but on a much larger scale. Balfour Stewart has suggested that, inasmuch as the Sun's brightness increases on the approach of a planet, and especially in that part nearest the planet (probably due to the reflection of light as well as to gravitative disturbance), these variable stars have large planets revolving around them at comparatively small distances, and that the brightest part of the surface is turned toward the planet. If the planet have a prolonged elliptical orbit of revolution, then the star will be very bright during only a small part of the period. And we may add to the theory of Stewart that if the axis major of the orbit has also a rapid movement, like that of the Moon, this fact will account for the differences exhibited in successive cycles, as the bright side will sometimes be turned toward us and sometimes from us, at the time of the actual maximum. It is not true, however, that the greatest brightness occurs at the time of nearest approach by the revolving body. The intensity of luminous disturbance appears to be rather due to planetary configuration than to mere distance.

Our Sun is a variable star; like Mira.

A remarkable example of another class of variable stars is found in Eta Argo, situated only 31° from the South pole; and therefore too far South to be seen in this latitude. It has a periodical change, between the 1st and 4th magnitudes, in a space of 46 years. And it is singular that a nebula surrounding it appears to be growing richer at the expense of the star; increasing in brightness as the star diminishes in lustre, and also undergoes a marked change in form.

A new star appeared suddenly in 1572, in the constellation Cassiopeia, and was visible for 17 months; its light being at one time so brilliant that it was equal to that of the brightest of the planets, and was visible at noonday. It suddenly disappeared, and was not afterward visible. But we find it recorded that in the years 1264 and 945 A.D. a similar appearance was noticed in the same region. These three dates are separated by intervals of 308 and 319 years; hence it is considered as probable that the star will re-appear before the year 1900. If so, this star presents us with a case of a variable of very long period, which is very bright at its maximum, and fades, even beyond telescopic vision, during the greater part of its cycle.

But there are some stars which seem to vary from other than mechanical causes. The most notable of these is one known as Tau in the Northern Crown. It was a telescopic star, of the 9th magnitude, previous to May, 1866. In that month it suddenly became visible to the naked eye; and on the 12th instant was nearly equal in brilliancy to a star of the 1st magnitude. On the 14th it was only of the 3rd magnitude; and it rapidly faded away to its original obscurity. Examination with the spectroscope led to the inference that this sudden apparition was caused by a genuine conflagration of hydrogen gas in the atmosphere of the star. Now; we know that immense aggregations of hydrogen are present in the Sun; and it is not impossible that a similar phenomenon has occurred in the past, or may occur in the future, in the centre of our system; though it is not probable, owing to the apparent absence of oxygen. What would be the result of such a change in the Sun? It would undoubtedly cause such an immense increase in the temperature of the Earth as to revolutionize our globe completely, killing off most of the plants and animals now existing, calling out mighty eruptions of the fluid interior, and leaving the world a wreck; perhaps to be peopled by new forms, adapted to the altered conditions of our planet. It is not the province of true science to attack or defend the truth which is revealed; but we can not resist the thought that possibly this is the way in which the Scripture will yet be fulfilled. A conflagration of hydrogen in the Sun, like that observed in Tau Coronæ, may be the agency chosen to cause the heavens to gather as a scroll, the elements to melt with fervent heat, the Sun and Moon to seem to turn to blood, and bring about the new firmament, and the new Earth whereon dwelleth righteousness.

We have no geological record of such a conflagration in the past; but we have abundant evidence of the fact that the minor changes in the condition of that variable star which we call the Sun, are not only caused by planetary motion, but have a reflex influence on planetary conditions. In my theory of the "Sun-spots and their Lessons," I have shown how the Earth was affected in its meteorological conditions by those phenomena in 1870; and the deductions have been accurately borne out by the observed facts. The spectroscope shows us that the dark rays emit comparatively little heat; but are prolific in the elements of chemical and electrical change. The result of Sun-spots last year was a reduction of about 2° in the amount of heat supplied to the whole globe—of earth, water, and air. This caused a reduction, of not less than four inches in depth, in the amount of water taken up from the ocean by the evaporating power of the Sun, and the necessary consequence of this diminished cloud supply was a diminished rain-fall all over the civilized globe; the season was one of the driest ever known, and the effects were wonderfully marked in the yield of the crops, the course of commerce, and the welfare of nations. Our corn crop was immense, as a consequence of the dry time, which almost totally suspended traffic on the Illinois river, and its connecting canal; and we can not say how much the course of events in Europe would have been changed, if the French gunboats had found water enough in the Rhine to enable them to operate against the Prussians. The effect of the diminished heat supply to the globe was not, however, a cold summer. The rain-fall being reduced nearly one-half, there was much less water to be evaporated from the land than usual — water taking up 1,100° of heat in passing from ordinary temperatures to a state of elastic vapor. Hence the slightly decreased heat received from the Sun actually produced a greater heat to the senses; not being carried off in the ordinary way into the upper regions of the atmosphere. The effects of this lack of evaporation are shown in the deserts on which rain never falls, because the cloud-bearing winds are intercepted by surrounding mountains. The thermometer often marks 130° in the Desert of Sahara, though the average temperature in that latitude is not more than 84°.

5

Changes of equal magnitude were doubtless caused in the other bodies of the system, during the past year, by the Sun-spots; and greater Sun-spot exhibitions in the past have undoubtedly wrought far greater changes on the surface of our planet than those which occurred in 1870.

Many of the fixed stars exhibit changes of color, without remarkable variation of brightness. Sirius was spoken of by the ancients as of a fiery red, but was recently white, and is now assuming the green color. Capella was also a red star formerly, afterward yellow, then white, and is now turning blue. These phenomena prove to us that extensive changes are in progress, the character of which we are as yet unable to determine.

These minor cycles of change, though great in themselves, are very small when compared with those indicated in the facts we have considered in this lecture. We have seen that the Moon was once like the Earth of to-day; and that both were once like the Sun, giving out light and heat as glowing masses of incandescent matter. We find nebulous aggregations which appear to be, as yet, uncondensed into the more solid form assumed by planets, Sun and stars: and we have discovered in all some of the very same chemical elements which make up a large proportion of the mass of our Earth. Philosophers have taken up these, and other, facts; and traced back the analogies presented, to what appear to be legitimate conclusions. We have not time to dwell on the successive steps of the reasoning process, but will present in brief the past history of the Universe, as it has been read out from present facts; just as the geologist has traced the past history of our own planet, from a study of the records that are deep graven in the rocks. They exist there in a language which though not alphabetic is universal, and can be understood by the men and women of every clime — into whatever tongue they may be obliged to translate those records in order to communicate the knowledge to their neighbors.

The space which we now call the Universe was once filled with matter in a state of extreme tenuity, fully three thousand million times less dense than atmospheric air at the Earth's surface, or two and a half million million times less dense than water. This was the original chaos. The matter was equally distributed, because no laws of aggregation existed. The establishment of the law of attraction of gravitation was the first great creative act.

If this law were first impressed upon a few adjacent atoms, they would at once move toward each other, and form a mass; if the same law were then instituted throughout the Universe, all the surrounding atoms would at once be attracted toward the first formed nucleus, the position of which would be preserved, because the attraction would be equal in every direction; and for the same reason the gathering particles would form a globe-shaped aggregation. The mutual friction of the atoms would generate the force that would cause heat as the first effect of the compression; then electrical excitement, as the atoms came more closely together; then light, as the vibrations of the mass were communicated through the partially cleared spaces outside the aggregation.

Millions of these centres and aggregations might be formed, through the immense void. We may take one of them in our mind's eye, and glance at its subsequent history. The inevitable tendency of the mutual flow of atoms would be to produce a whirling motion of the mass; as the flow of water in a funnel is always accompanied by a rotary movement. The attraction still operating, the interior atoms would be forced nearer to each other by the pressure from without; but this would be partially counteracted in the equatorial portions, by the rotary movement, and the combination of forces would produce an ellipsoidal mass—flattened at the poles in exact proportion to the velocity of rotation on the axis. These changes would continue till an equilibrium was established; the centrifugal force at the surface of the equatorial part being equal to the force of attraction toward the centre.

During all this time the action of light and heat, on the particles of matter, would aggregate them into smaller groups, forming chemical atoms, possessing different properties; and then would begin the play of chemical affinities. These secondary atoms, combining in various proportions, would form definite substances, giving rise to subordinate centres of attraction, about which the adjacant atoms would gather. The movement toward the new centres would produce a rotation of each mass, as in the first instance; while the motion around the original centre would continue, becoming a revolution in the case of each individualized mass. The tendency of the interior portions of the original aggregation, to move with greater angular velocity than those farther from its centre, would lead to its separation into rings, or bands. The separate parts of each ring might aggregate into one planet, as in the case of Jupiter and Mars, or into several bodies, as in the case of the planetoid family revolving between the orbits of those two greater planets; some of these rings doubtless exist still, without aggregations large enough to be visible to us, except when they come within the limits of our atmosphere.

The same process of throwing off rings might be repeated in the case of each planetary mass; and these secondary rings would either revolve, as such, around the planet—as is now the case with the rings of Saturn—or form smaller bodies by aggregation—as in the case of our Moon, and the satellites of Jupiter. The liability to this division would be in proportion to the magnitude, and the consequent difference between the angular velocities of the interior and exterior portions—the actual velocities being nearly equal, as being produced by the same moving impulse toward the centre. Accordingly we find that the largest planets have the greatest velocity of axial rotation, and several moons; the smaller ones have none at all. But each *may* be a centre of revolution to millions of bodies too small to be discerned; only the smaller planetoids being actually destitute of attendants.

The Earth, as thus formed, would be, at first, a much larger mass than at present; her circumference extending far beyond the present orbit of the Moon. For millions of years after our satellite was thrown off, the Earth would glow at more than a white heat, as a result of the mutual compression of its component matter. It would cool rapidly at the surface; the lighter particles taking the outer position, by virtue of their less specific gravity, and forming an atmosphere of vapor. Through a long course of ages the cooling process would continue, the surface gradually hardening into solid matter, but becoming fissured by contraction, and falling into the fiery sea, whose commotions lifted up other portions of the crust, forming continents and islands. As the irregular surface became cooler, the vapors in the atmosphere would condense into water, forming seas; and then the air and water, acting on the surface of the granitic crystal, would gradually wear away the material from which the aqueous rocks were afterward formed, and a foothold produced for the first forms of vegetable existence. The subsequent part of our wonderful Earth history is written in the rocks; it is the province of the geologist to trace it out.

Thus, all the visible and invisible objects in the Universe were formed from the one homogeneous material.

Till recently it was thought that the regions of space were swept clean by the attractions of those greater aggregations which we can see; but we now know that what is called space is yet comparatively full of material particles. Unnumbered millions of aggregated atoms still course about within the bounds of the solar system; and besides these the whole space is full of unattracted matter, yet unformed into even the smallest masses, which sustains much the same relation to the planetoids as the air does to the motes that float in it, as revealed by the presence of a sunbeam. There is an old saying, to the effect that "Nature abhors a vacuum." It would probably be impossible to exhaust the interplanetary spaces; just as it is impossible, by the most perfect air pump ever invented, to form a complete vacuum on the Earth's surface.

Indeed; it is probable that this unformed matter, the existence of which has been already demonstrated by the shortening of the period of Encke's comet to the amount of three days within the past eighty years, and which doubtless acts in the same way on the more dense planets, but to an extent not yet appreciated, has other, and very important functions. The diffusion of Sunlight in the atmosphere has recently been shown to be directly due to the presence of the motes floating in the aerial envelope of our globe; as I announced in March, 1870, several months before the fact was stated by Professor Tyndall. And there is reason to believe that the same office is performed in the interplanetary spaces by the asteroidal matter, while the propagation of the ray in a right line is due to the presence of the ether, or unformed matter; and it will yet be proved that not a single ray of absolutely unpolarized light ever reaches our eyes. The theory of the polarization of light requires some modification to make it harmonize with the facts.

This ether, and these planetoidal masses, whether specks or lumps of matter, are all composed of the very same element, or elements, as those which make up the sum of the planets and the stars. It is only the same matter, existing under different conditions; and it is idle to dream, as some have done, of the existence of an ether, like that of the interplanetary spaces, existing between earth atoms, to account for the phenomena of heat. All that we know of the constitution of matter tends to indicate that there is really but one original element, though we enumerate sixty-four: and I believe it will yet be discovered that the diverse elementary atoms differ only in this, that each class of atoms has its own capacity for specific heat, being formed of groupings of the original particles in such a way that differing quantities of heat excitation are required to produce the same amount of heat-movement in the atom How this property was communicated we may never know; but it is not difficult to conceive it as the result of atomic formation at different stages in the cooling process. The atoms of each of the so-called chemical elements have a separate rate of thermal vibration; and the admission of this theory affords an ample reason for the display of diverse properties. It will be remembered that each element gives lines in the spectrum which are only distinguished from the rest by specific lengths, and rapidities, in the arcs of vibration. We may add, too, that this theory will enable us to account for the differences of atomic constitution in bodies formed from the same mass; and will also obviate the objections recently made to the igneous origin of the primary rocks.

*

We can not help being struck by the fact that the Universe presents to our view almost every different stage of growth and decay. The gaseous accumulations of matter, which we call nebulæ, are in the earlier periods of existence — unborn suns. The Sun and stars are cooling bodies; and will require millions of years to enable them to reach conditions similar to those of the Earth to-day. The cometary orbits lose a portion of their ellipticity with every revolution, and will settle down into staid members of the planetary family in the far distant future. The Moon has passed through the stage of active existence, and is going through her period of decay. The planetoids, which are continually falling to the Earth, and to the other planets, are bodies whose independent missions have been accomplished; they have yielded up their separate identities, and gone to form portions of other and larger worlds. Pursuing this analogy to its legitimate sequel, we see, in the vast future, suns losing their light, and planets dying out from cold: we see moons falling to planets, and planets falling to suns; and suns, again, subsiding into larger aggregations of matter. All this must occur in obedience to the law of attraction of gravitation; because that force is constant, while its counterpoise, heat, is indubitably susceptible of decrease by radiation. We may go further. We may say that heat, itself, is but the effect of gravitation; and that, though the effect is at present equal to the cause, yet the cause is greater than the effect, in this respect, that the cause is perpetual, while the effect is evanescent. All the phenomena of motion, of heat, of light, of life, are traceable to the operation of this one great, incomprehensible force which we call attraction. All the phenomena must ultimately cease, by virtue of the operation of this force; as the forces that produce and sustain human life for a brief period, are themselves the real causes of that which we call natural death.

It is even possible to wing the flight of human thought over, and beyond, the countless stretch of ages which will measure the duration of the visible Universe. We can not absolutely know, but we may reason out from analogy, the idea, that a new Universe may be evolved from the general wreck; and that without a new exercise of Creative energy. The concussion of such immense masses of matter as those into which the penultimate cosmos will be resolved, may generate heat enough to redistribute the vast concourse of matter into its former diffused chaos, and permit the processes of condensation and formation to commence anew. And indeed, we know not but that our Universe may be only one of a vast series of such successive creations—a far down link in the mighty chain of involution and evolution of material forms and forces, which is eternal to our finite minds, inasmuch as we can not even guess at its possible limitations; though it be still subordinate, in function and duration, to the Great First Cause, whom men call God. And let not this thought be deemed impious! We sin not in endeavoring to grasp a scintillation of the Omniscience which beams on the throne of the Eternal; we blaspheme only when we assume to sit on that terrible throne as a judgment seat, and dare to say that He hath not ordered all things well.

Man has been called the microcosm — an epitome of the Universe. The individual who first gave shape to that thought was probably unable to comprehend its full scope. In a mental relation, man is the image of his Creator; physically considered, he is a type of the Universe, through all the phases of growth, maturity, decline and extinction. Like him, the Great Universe is permanently impermanent. But the same is true of *every* part of the great whole. The plant gathers material from the soil and the atmosphere, pushes forward into life, attains to maturity, withers, dies, and rots away into thin air. The human body passes through a similar consecution of growth and decay — of formation and dissolution. Families of men have *their* periods of existence, but of longer duration; nations pass through the same phases, but occupy still longer periods in the history of their distinct existence. As it is with nations, so with types of existence; and as with types, so with the conditions under which they exist. These conditions, again, form but items in the history of a world formation, as the rotation of a planet on its axis is an item in the journey around the Sun. So the history of a world is but a part, and an epitome, in the history of a system; and *that* of a Universe. In all we see the signs of constant growth, up to certain limits; then a staid maturity; then an old age; then death. It is true of the Universe as of man — we all do fade as a leaf. All wax old as doth a garment. Only one being exists of whom it can be said:

> "Thou art eternally the same,
> Thy years shall not fail."

I have been asked: How do the conclusions stated in these lectures, in reference to the creation and dissolution of the Universe, and other deductions made by scientific men, harmonize with the teachings of the Bible? and, if there be a disagreement, how shall we decide? I answer:

The two appear to disagree, in many important respects; and a storm of controversy on the subject has raged in the Christian world, with a few occasional lulls, ever since Copernicus announced that the Earth is not the centre of motion of the visible Universe. But it is worthy of remark that the debate was far more violent and bitter a few score of years ago than it is to-day; and its violence abates in exact proportion to the progress of man in the study of the physical sciences, and in a critical knowledge of Bible meanings. The more we know, the less of disagreement do we find between the teachings of the Bible and the results of scientific research.

All truth is consistent with itself; but if we find two assumed truths which appear to be inconsistent, it is no more probable that one of the assumptions is false, than that we have failed to understand one or both of them.

It is also true that differences of opinion arise much oftener from a want of agreement as to the precise meanings that should be attached to different statements, than on account of a radical disagreement in the statements themselves.

The generally received acceptations of the statements made in the first chapter of Genesis, do not harmonize with our geological theories. But that fact does not warrant the infidel conclusion that the Bible is an untrue record; neither should it prevent us from studying the book of Nature. There is a mistake somewhere; but it probably lies with those who first make up their minds that the Bible teaches this or that, and then anathematize all who will not accept their interpretation of its language.

The Bible was evidently not intended to be a teacher of physical science. The references made in that volume, to physical facts, are couched in the idioms best understood by the peoples who lived in the times and places where its several books were written. Even to-day, WE speak of sunrise and sunset, though we know very well that the Sun does not move to cause either of those phenomena. No one understands the Biblical allusions to "the four corners of the Earth," or its "firm foundations, that can not be moved," in any other way than as popular phraseology. Similarly no well-informed person now believes that "the Sun stood still upon Gibeon," at the command (or prayer) of Joshua, though he may accept the miracle of an extended day. And, yet, we have just as much warrant in the Bible for assuming that the Earth is the centre of the solar system, as we have for believing that the Earth, Moon, Sun, and Stars, were all created, out of nothing, in six days of twenty-four hours each. Indeed, Moses does not tell us that matter was formed from nothing; he simply says that, "In the beginning * * * the earth was without form, and void"—which is just what science teaches to-day. To create is to fashion, to shape, out of previously existing material.

A great proportion of the creed called "Christian" is falsely assumed to be found in the Bible; it really originated in confessedly human productions — from the writings of the Fathers, down to Milton's Paradise Lost. We can not find within the covers of the Bible a hint that the sacred volume was intended to be a teacher of Science, though we do find its uses defined. St. Paul tells Timothy that the Scripture is "profitable for doctrine, for reproof, for correction, and for instruction in righteousness; that the man of God may be perfect, (and) thoroughly furnished unto all good works." In the two sections of this inspired volume we find only two great fundamental truths, a belief in which was insisted on; the Old Testament anathematizes those who deny or forget the God of Israel; the New Testament teaches that Jesus is the Christ. The latter truth was the only one to which the Eunuch gave assent as a preliminary to his baptism by Philip. Indeed it is well for the Christian world that the prophecies are the only portions of the book on which the Scriptures themselves forbid us to put a private interpretation. If it be a sin for scientists to dispute the alleged meanings of a few Scriptural passages, which have no important bearing on Christian faith or a godly life, then there is not a church in existence but is a hundred times more guilty of the crime of "wresting the Scriptures"; and the verdict of guilty would be pronounced against every one of them, by the unanimous voice of all the rest.

It may be claimed that the philosophers of our

age are no better agreed than the churches; that the knowledge of to-morrow may largely modify many of the deductions of to-day, as the researches of this generation have demolished much that was accepted as truth in the past. We admit this; and claim that the fact is the best guarantee of ultimate wisdom as a result of honest, fearless investigation. We are sure of very few facts or truths, yet, in comparison with the great mass of facts which yet remain to be mastered, and of truths yet to be patiently reasoned out. The present state of scientific knowledge may be compared to the primary state of ossification in the unborn child. Bone commences to form at several detached points, and the growth has proceeded to a considerable extent before the extensions from these centres meet each other, and give the true skeleton. So, we know a great many isolated facts, and have reasoned out a few principles; but we are far from having reached the stage of development where we can take a bird's-eye view of things, and trace out their mutual connections and intercausations, as "parts of one harmonious whole; whose body Nature is, and God the soul."

In the book of revelation, and the book of Nature, we have two works which we believe to be the productions of one Author. For both we may claim that the difficulty of understanding them is the best proof of their divine origin.

Any mere human work would have been mastered in less than one thousandth part of the time that has been spent in the attempt to comprehend all that is written in either of those two volumes. The hieroglyphs of India, Egypt, and Assyria, of which the characters and the language were alike unknown, have been read out within the past hundred years; whereas these Divine books contain many thousands of mysteries yet undeciphered, though they have been studied for scores of centuries.

Believing this, we can but consider it our duty to study both; and our high privilege to be permitted to do so. The intense mathemaphobia exhibited by some theologians can only be accounted for by supposing that they believe the book of Nature to have been written by the father of lies, for the express purpose of leading men astray.

We may feel assured of this: He who created all has made nothing in vain; nothing that he did not intend to be used; not a single talent that was designed to be buried in a napkin. The gift of reason was conferred on man that he might study the Great Author of the Universe in His works, and build himself up into a closer mental likeness of Him who knoweth all things. We have not yet progressed far enough in either department of the investigation to be able to make a dogmatic comparison of notes. But we are justified in believing that the Source of all Truth can not be inconsistent with Himself; and that, if we are ever able to grasp the whole truth of Revelation and of Nature, we shall see that all is in perfect harmony—just as we now know that the two opposing forces, known as the centripetal and the centrifugal, originated in the one force which we call the Attraction of Gravitation.

NOTES.

Pages 17 and 105. The number of chemical elements absolutely known is sixty-four. The existence of three or four others is supposed rather than ascertained.

Page 20, line 26; for *this* solar spectrum, read *the* solar spectrum.

Page 31. The calcium light is the most intense we can produce, except the electric light. The limit of the latter is not known. The electric spark has been made to give a light nearly sixty times more intense than the calcium, or only two and a half times less intense than the sun light.

Page 37. The chemical action of the sunbeam is now believed to be principally due to the presence of chromium, titanium, and magnesium in the sun.

ANNOUNCEMENT.

I HAVE now, nearly completed, the material for a work on Astronomy, of which the following are the prominent features:

Vol. 1. Full and comprehensive star maps, engraved in the very best style, and in such portable form as to be readily compared with their originals in the sky. Also, lists of interesting telescopic objects, with appropriate illustrations and full descriptions. The third page of the cover contains a specimen of the star maps, which will, however, be printed on first-class paper.

Vol. 2. The methods of ascertaining the distances and dimensions of the heavenly bodies, explained so that everybody can understand the processes. The laws of world motion, similarly simplified, and the movements of earth, moon, planets, and stars explained, with the resulting phenomena. Star chemistry, and the biology of the universe, discussed somewhat more fully than in the preceding pages.

Everything will be brought down to the latest date, and reduced to the simplest possible form; so that he who runs may read, and all who read may understand.

I wish to produce a work which will be a credit to the West, and to this end propose to publish it by subscription. If I obtain the names of subscribers for five hundred copies I will put the book to press. If the required encouragement be not forthcoming, I shall conclude the book is not wanted.

I, therefore, invite orders, for one or more copies, from those interested in furthering so important a work, promising to pay the retail price, not exceeding ten dollars per copy, on delivery. In return I promise that it shall be worth the money.

E. COLBERT.

Address, Chicago *Tribune.*

www.ingramcontent.com/pod-product-compliance
Lightning Source LLC
Chambersburg PA
CBHW021821190326
41518CB00007B/695
9783337042356